Introduction to
Appropriate
Technology

Introduction to
Appropriate Technology
Toward a Simpler Life-Style

Edited by R. J. Congdon

 RODALE PRESS, Emmaus, PA

Printed in the United States of America on recycled paper

2 4 6 8 10 9 7 5 3 1

Library of Congress Cataloging in Publication Data
Main entry under title:

Introduction to appropriate technology.

  Bibliography: p.
  Includes index.
  1. Technology. I. Congdon, R. J.
T49.5.I58   338.91′172′4   77–10596
ISBN 0–87857–188–4

Table of Contents

Foreword

This book was first published by the Committee for International Cooperation Activities of the Technical University of Eindhoven in the Netherlands, as a result of a series of lectures organized jointly with the CICA at the Technical University of Twente, in the Netherlands. The lectures were given at both universities at the end of 1974, and were part of the growth of activities in the field of socially appropriate technology in the Netherlands which the two CICA's have done much to stimulate.

The first edition of this book was published by the TOOL Foundation, a nonprofit agency that coordinates the research and development activities of volunteer groups of staff and students in Dutch higher education institutions, and an engineering consultancy firm. Since its establishment in 1974, TOOL has handled thousands of technical inquiries from both industrialized countries and the Third Worlds, from development workers engaged in activities in the field of socially appropriate technology. In addition, through the medium of publications such as this, TOOL has promoted the concept and the practice of a socially appropriate technology both to the general public and to current policymakers. Engaged in providing support for a growing number of field projects in Africa, Asia, and Latin America, TOOL is working with government and nongovernmental agencies alike on the application of SAT. A key activity in which TOOL has been encouraged by similar agencies elsewhere to take a lead is in the establishment of a comprehensive documentation bank for the practitioners in SAT; SATIS, the socially appropriate technology information system, is an informal and practical network of development workers spread throughout the world.

Neither the Technical University of Eindhoven, nor the TOOL Foundation, can accept any liability for damage claims arising from the use of any advice or descriptions of any instruments or apparatus contained within *Introduction to Appropriate Technology*.

R. J. Congdon

Introduction

Between the bullock cart and the jet plane, between the hand plow and the combine, between the wheelbarrow and the truck, there exists a whole range of "intermediate" technologies that are "appropriate" to an individual, a culture, a nation.

In this age of big, and often destructive, technologies, increased attention is being given to "intermediate" or "appropriate" or "low-cost" technologies which tend to mean different things to different people. But, beyond this semantic tangle is a discernible cry, both in America and abroad, for alternate technologies, for different "world views" and tools and technologies to fit such perceptions.

In his book, *Small Is Beautiful*, E. F. Schumacher coined the term "intermediate technology" to describe a "technology of production by the masses, making use of the best of modern knowledge and experience, conducive to decentralization, compatible with laws of ecology, gentle in its use of scarce resources, and designed to serve the human person instead of making him the servant of machines."

From an economic point of view Schumacher considered intermediate technology to stand between high and low technologies. "I have named it," he wrote, "intermediate technology to signify that it is vastly superior to the primitive technology of bygone ages but at the same time much simpler, cheaper, and freer than the super-technologies of the rich. One can call it self-help technology or democratic or people's technology—a technology to which everyone can gain admittance and which is not reserved to those already rich and powerful."

In a specific sense intermediate technology represents a practical alternative to those caught in the grips of "inappropriate" technology. In his contribution to this book Peter D. Dunn refers to the inappropriateness of a European space agency supplying television sets to rural parts of Africa and then supplying solar cells, battery packs, and small gasoline engines to supply the power for the sets. "Gasoline engines," Dunn suggests, "are unreliable and difficult to maintain, battery packs are very expensive, and solar cells are prohibitively expensive. Now this is a good example of where pedal power is required. A person can pedal very easily on a bicycle for an hour and generate 35 watts. We have in fact developed a generator of this sort. It consists of a small wooden frame with a wheel-mounted dynamo. If it goes wrong, the reason is obvious; the dynamo has fallen off or something similar. The local people do not regard it as a mysterious black box. They see what is wrong and they can repair it themselves and make more units at the same time."

But whether a technology is "intermediate" or not depends on a number of factors. The ox-plow, which stands halfway between the traditional hand-operated hoe and modern diesel tractor, would be considered an intermediate technology in Africa where it was recently introduced. On the other hand, writes Nicolas Jéquier in *Appropriate Technology: Problems and Promises*, "In the societies of the Middle East and Asia which have known and used the ox-drawn plough for thousands of years, such a technology can be called traditional, and the intermediate level of technology would more adequately be represented by the small two-wheel tractors of the type developed by the International Rice Research Institute in the Phillipines or by the industrial co-operatives of Sri Lanka. In the tropical African societies which do not have any tradition of livestock breeding and which still use very simple implements, the ox-drawn plough is a major innovation, and from a technological point of view, it represents a big step forward."

Therefore, in the very best sense, the "intermediateness" of a technology is relative to time and space, to the perceptions of a particular culture, and to the kind of engineering it has enjoyed. And, by narrow definition, "intermediate technology" belongs to the field of engineering.

Because of the economic and engineering overtones of "intermediate" technology, people seem more comfortable with the term, "appropriate" technology which represents, among other things, the social and cultural dimensions of this movement. In addition, appropriate technology is a vehicle for certain positive and symbolic insights. While appropriate technology refers to the middle-level workplace, it also refers to a kind of social and cultural revolution that is evident in many parts of the world. According to Ken Darrow in *Appropriate Technology Sourcebook* this "technology is especially attractive because it seems to solve a number of problems at once. Because it involves self-reliance and local production for local needs, on a national level this approach can remove from the list of obstacles to development many of the inequities of an international system that is dominated by the expensive technology and economic power of the rich countries. At the same time, the lack of well-developed infrastructure and the shortage of highly trained manpower to run large industrial operations become much less important when people are allowed and encouraged to develop themselves wherever they are. A whole array of problems can potentially be solved at once.

"It is precisely for these reasons that the appropriate technology concept is spreading in popularity so rapidly. Those who believe in small entrepreneurial capitalism, decentralist Marxism, European socialism, African communalism, and Buddhism can all find much of value in the ideas underlying appropriate technology. Different people are attracted to it because it seems to address so many fundamental problems so directly."

Accordingly, appropriate technology may best be seen as a movement, a humanistic counterweight to the mechanistic view of the world that has prevailed for the last few centuries; as an opportunity for all citizens of the world to participate in new styles of architecture, a change of heart.

On the other hand, appropriate technology may refer to specific lists of tools and techniques, which, as listed in *Appropriate Technology Sourcebook*, share the following characteristics:

1. Low in capital costs
2. Use local materials whenever possible
3. Create jobs, employing local skills and labor
4. Are small enough in scale to be affordable by a small group of farmers

5. Can be understood, controlled, and maintained by villagers wherever possible without a high level of Western-style education

6. Can be produced out of a small metal-working shop, if not in a village itself

7. Suppose that people can and will work together to collectively bring improvements to the communities, recognizing that in most of the world, important decisions are made by groups rather than by individuals

8. Involve decentralized renewable resources, such as wind power, solar energy, water power, methane gas, animal power, and pedal power (such as in that highly efficient machine, the bicycle)

9. Make technology understandable to the people who are using it and thus suggest ideas that could be used in further innovations

10. Are flexible so that they can continue to be used or adapted to fit changing circumstances

11. Do not involve patents, royalties, consulting fees, import duties, shipping charges, or financial wizards.

Perhaps appropriate technology can be seen as an evolving book of people-oriented ideas, tools, and attitudes which both the developed and developing countries contribute to in various ways. Undoubtedly appropriate technology, which can be seen as a kind of probe in the industrialized countries, has its modern roots in the developing world.

Introduction to Appropriate Technology

Introduction to Appropriate Technology, first published by the Committee for International Cooperation Activities of the Technical University of Eindhoven in the Netherlands as a result of a series of lectures organized jointly with the CICA in the Technical University of Twente in the Netherlands, is a comprehensive collection of essays by eminent economists, sociologists, engineers, and others on appropriate technology.

Very simply, this is a revolutionary work by "technical revolutionaries," many of whom have been associated with the Intermediate Technology Development Group, who believe that appropriate technology offers genuine alternatives for both developed and developing countries. Moreover, *Introduction* is a lively critique of Western technology that, in the words of Ton de Wilde, "can only aim for a society where people are consumers and no longer producers. In our Western rationality, in our dialectic way of thinking, we distinguish between consumers and producers, between life and work, between capital and labor, between materialistic things and spiritual things or feelings. But what happens when the decision makers base their decisions on models that only recognize a part of our human being?

"Now, on the one hand we have our cars— sometimes with stereo—our color television, our household gadgets, our parties; on the other hand we have our psychiatric hospitals. And there are simply not enough of them to cope. We do not have enough psychiatrists and psychologists to help the people who run their heads against the wall—people who want to work but can only go each week to the social service to be paid for doing nothing. It must be clear by now that to me technology means the labor situation and the whole technological infrastructure.

"The technological network is concerned with a mass of information, the speed of life, places crowded with cars, a lot of garbage in nature, the spoiling of our ecological system. For some people in some situations this sort of system has its benefits, but we must realize that there is also a bad side for other people in other situations. This is true not only at a national, but also at a world level. One part—the 20 percent in the developed countries—is well fed and has some kind of work, while the 80 percent in the less developed countries is badly fed, and there is no work for 30 to 50 percent of them."

Introduction to Appropriate Technology is an important, perhaps essential, handbook of tools and techniques for rethinking development aid. Moreover, it provides specific rationale and content for appropriate technology; it is full of the hardware of appropriate technology. For example, John P. Parry concludes his chapter in "Intermediate Technology Building" with the following admonition: "When the people in the developing world can build themselves good brick and tile houses with the masonry bonded by pozzolanic mortars, all of which materials they have made themselves, we will have taken a major step towards reattaining the standard of maintenance, insulation, and permanence of that excellent but scarce dwelling, the Ice Age

cave." And such a hope is also applicable to residents of the American Southwest who, for these reasons, are rediscovering a traditional and worthy building material: adobe.

To those critics who suggest that appropriate technology is a loose collection of ideas and inventions, *Introduction* gives lie. Appropriate technology, as the following chapter titles indicate, is applicable in a multitude of situations: "The Social Context for Choosing Water Technologies," "Tools for Agriculture," "Intermediate Technology Building," "Energy in Rural Areas," "Pedal Power," "Intermediate Chemical Technology," "Education Systems," "Appropriate Production Systems," Industrial Liaison," "The Transfer of Knowledge and the Adoption of Technologies."

But more than exploring the hardware of appropriate technology, *Introduction* explores the spirit of this movement with examples from revolutionary societies, such as Cuba, China, and North Vietnam, where people-oriented technologies are a way of life. In chapter 12 "The Transfer of Knowledge . . ." Harry Dickinson notes that "China is feeling the effects of mechanization. If you ask the people in a village what happens when they get a pump, they say that they all work less hard. But that is only their immediate reaction to the innovation; after a while they say that they start up new 'sideline' industries. These may give them the added value from processing their own raw materials; they may produce goods that the community itself needs, or they may produce goods for sale and thus enable the community to purchase products it was not used to before. Whatever the objective of the new activities, there is no idea that anything to do with increased productivity could lead to any sort of unemployment. Such a result is inconceivable."

Appropriate Technology in the United States

There is an abiding belief on the part of scores of groups in the United States that appropriate technology offers real promise for a kind of cultural revolution and a basis for genuine brotherhood. This thinking is reflected in the report, "Appropriate Technology in the United States," prepared by Eugene Eccli for the National Science Foundation. "The problems," Eccli writes, "faced by appropriate technology innovators and a world moving toward a recycling-based economy are not insoluble. A high degree of initiative and responsible local activity, coupled with the opening of economic opportunities and quality control gained by a people-to-people communications system, offer a chance to build a matrix of local solutions whose combined effect would address critical issues like new enterprise development, meaningful employment, and environmental renewal. In the worldwide context, appropriate technology could provide a basis for cooperation between small innovative groups in the United States and developing countries. Moreover, as its efficient methods are adopted it could help to relieve international tensions caused by resource shortages.

Admittedly, it is somewhat paradoxical to think of appropriate technology serving the needs of both developing and developed countries. And it is: while, for example, parts of Africa struggle to enter the industrial age, parts of America struggle to enter the post-industrial age. Nonetheless, in spite of the vast difference in needs among the populations, appropriate technology can be a two-way street, a vehicle for understanding. As Eccli notes in his study, "The U.S. can learn from Japan, Egypt, Taiwan, and Israel, countries whose small farmers contribute greatly to food production and who have mastered many of the techniques of rural development. Of particular interest are the institutions which protect small farmers in these countries, keeping them involved in highly productive agriculture, and the relationship between those institutions, appropriate technology, and economic incentives."

National Center for Appropriate Technology

Although appropriate technology surely addresses pressing needs in developing countries, it does have relevance to industrial societies, such as the United States, which recently formed a National Center for Appropriate Technology to develop and implement technologies appropriate for low-income communities. From a historical point of view, this national interest in appropriate technology is similar to the environmental and consumer protection movements; it grew out of a specific set of circumstances, including the energy shortage, and not particularly out of events in the Third World. Still, the National Center for Appropriate Technology should strike a cord with appropriate technologists overseas. According to the NCAT, "*The main goal of appropriate technology is to enhance the self-reliance of people on a local level.*"

The NCAT, which will fund and advise individuals and communities on the basis of their particular needs, will be "hardware oriented," as the following objectives indicate:

1. Improving heat-source efficiency (e.g., improved flue dampers)
2. Low-energy cooling and ventilation
3. Reducing the cost of insulation through manufacturing techniques
4. Weatherizing mobile homes
5. Developing low-cost housing for rural areas based on indigenous and recycled materials
6. Developing community greenhouses
7. Developing aquaculture systems for low-income communities
8. Developing composting toilets and methane digestors for rural areas
9. Developing community industries based on the manufacture of small-scale technologies for community and outside use (such as solar collectors).

In truth, the establishment of the National Center for Appropriate Technology was hastened by the good work of existing appropriate technology organizations in America and Canada. These groups, including New Alchemy Institute, Zomeworks, Windworks, Institute for Self-Reliance, The Farallones Institute, Ecotope, the Brace Institute, Rodale Resources, and many others are developing appropriate-technology hardware and strategies.

Interestingly, the budding interest in America in small-scale technology would not have surprised the country a century ago. As Nicolas Jéquier notes "virtually all the industries which grew up in the United States in the nineteenth century started on a very small scale, often as one-man operations.

"Another lesson from the American experience is that, contrary to what happened in most European countries, a high proportion of the inventors and entrepreneurs came from the rural communities. Oliver Evans, the inventor of the automatic milling machine, was brought up on a Delaware farm; Eli Whitney, who was to play a crucial part in the development of the textile industry, and later the machine industry, grew up to manhood on his farm in Connecticut; Cyrus McCormick, whose name became the major trademark in agricultural machinery, was also a farmer's son, and Henry Ford himself came from a Michigan farm. Clearly, the American farming community of the nineteenth century was very different from the peasant societies of any other countries: the farmers were free men and they knew that the future would be what they wanted it to be. These examples are given here to suggest that development is not necessarily an exclusively urban phenomenon and that the inventiveness and the entrepreneurship in the rural sector are extremely important. This point must be emphasized, since more than 70 percent of the world population today still lives in rural communities. No society can be considered as truly 'developed' unless it has a healthy agriculture, and the social and economic level of the agriculture sector is generally a good indicator of a country's overall level of development."

In *Small Is Beautiful* Schumacher noted that the "applicability of intermediate technology is extremely wide, even if not universal, and will be obvious to anyone who takes the trouble to look for its actual applications today. Examples can be found in every developing country and, indeed, in the advanced countries as well. What, then, is missing? It is simply that the brave and able practioners of intermediate technology do not support one another, and cannot be of assistance to those who want to follow a similar road but do not know how to get started."

It is hoped that *Introduction to Appropriate Technology* will be of great help to both those who "do not know how to get started" and all those who are interested in the fullest application of appropriate technology at home and abroad.

James C. McCullagh

Chapter One

An Approach for Appropriate Technologists

by George McRobie

In recent years there has been considerable effort in the area of what might be called people-oriented technologies, which have been variously called intermediate, appropriate, progressive, and low cost.

By virtue of the work done by the Intermediate Technology Development Group in London, the phrase "intermediate technology" has gained considerable currency. Invented by E. F. Schumacher after a visit to India in 1963, "intermediate technology" is essentially an economist's concept reflecting an alternative to the very costly technologies of Europe and America. On the average, in Western Europe technology costs approximately $6,000 per workplace. This figure represents the average cost of creating one workplace in a manufacturing industry, excluding the cost of land and building. Thus, it would cost $6,000 for equipment for the workplace.

Traditional technology, which is now in use in the rural areas of developing countries, costs about $2. Therefore, in the

Photo 1.1. Simplest technology—field workers in China.

1

Photo 1.3. Peasants weeding in India.

Photo 1.2. Trimming white potatoes for starch in Taiwan.

simplest sense intermediate technology refers to the cost of a workplace between high and low technologies.

Since the average income of skilled workers in the rich countries is $4,000, society has little difficulty in regenerating workplaces at a high level of cost. For example, a skilled workman who saves one month's salary each year can buy his own workplace in 18 years. But if his average income is only $40, he would be saving for his workshop for nearly 200 years.

Considered from this point of view, Western technology can spread only to a very small section of the population. And this is precisely what has happened. Development has been concentrated in cities; it has bypassed the great majority of people in the rural areas, and it has caused divisions in a society which are far greater than those which might have occurred without the advent of such a technology.

Infrastructure

The factory in the rich country is really the outward manifestation of a whole set of societal forces. It involves very specialized education, highly specialized transport systems—and the assumption that they are always inexpensive—mass markets, specialized fuels, and a very disciplined labor force.

Merely to take this factory and transfer it to a developing country does not automatically create the infrastructure on which its success depends. Not surprisingly, large-scale, capital-intensive industry in developing countries generally needs a great deal of support, and even then, usually runs well below capacity.

The challenge was this: How do we interest people in developing (and developed) technologies that are somewhere between $40 and $400 per workplace? Then, how do we determine what the characteristics of such technologies would be?

Photo 1.5. An animal-drawn, two-bottom moldboard plow (Kenya).

Photo 1.4. Intermediate technology—a seed planter.

3

Photo 1.6. Brickmaking in China.

First of all, we can be certain that intermediate technologies would be the opposite of what we have now. They would tend to be small, they would be relatively cheap, they would be low in capital cost, they would be relatively unsophisticated, and importantly, they would be non-violent, particularly in the sense that they would be largely based on indigenous resources and used for local needs. This last point distinguishes them from our sophisticated technologies which are based on the principles of a robber-economy, which strips resources from the face of the earth at maximum speed, turns them into short-lived consumer goods and sells them back at very high prices to the countries from which we originally got the raw materials.

Secondly, intermediate technologies would tend to work *with* nature rather than against it. It is a historical fact that modern technologies and economics have contributed to man's alienation from himself, his work, and from nature in general. However, intermediate technologies, the hallmarks of which are smallness, capital cheapness, and simplicity, would assure that the great majority of people should be able to actively participate in the conduct of their own lives.

Large-scale technologies encourage an economic exclusion. The real decision about how things are produced, where they are produced, who produces them, and who receives income from the production, is concentrated in a relatively small number of hands.

For these reasons we are asking engineers, scientists, and economists to look at technology through a new pair of spectacles. "Instead of concentrating on labor-saving devices, which has been the whole trend of modern technology, can you turn your attention to capital-saving devices, because it is capital that is lacking in developing countries, not labor." And

Photo 1.7. Advanced technology—power-operated thresher for small farms.

Photo 1.9. Tractor-mounted row planter, with fertilizer attachment (India).

Photo 1.8. Rototilling in China.

indeed, we may soon find this to be the case in many industrialized countries.

Intermediate Technology Development Group

On the basis of this perceived need the Intermediate Technology Development Group was formed in 1966, initially serving as a pressure group trying to change people's attitudes about what constituted aid. Within the group are various panels, each of which consists of the best technical experts we can find in the respective areas. The panel covers subjects such as agricultural tools and equipment, water supply, rural health, building and building materials, food technology, cooperatives, forestry, and chemistry and chemical engineering.

Methodology

The first thing ITDG did was to conduct a survey of what already exists within the categories of intermediate or appropriate technologies. From this we tried to identify certain gaps in knowledge about technology which need filling. Thirdly, we hope to mobilize technical knowledge to fill in these gaps. Fourthly, we try to disseminate this knowledge in a practical form.

Now while this might be very logical, it is extremely difficult to carry out. Consider the type of knowledge that is reaching developing countries. In practice most institutional forces are directed toward giving these countries only one type of knowledge—the knowledge of the most sophisticated technology that the rich countries have to offer. Such a process is not only built into our official aid programs, it is also built into the educational systems which the developing countries

have inherited or copied from us and which we continue to foster by means of training scientists, engineers, and administrators who return to their own countries with "inappropriate" technologies.

So at both ends—both the giving and the receiving—people's minds are channelled almost exclusively toward the "boundaries of knowledge" that industrialized countries trade in. If you look below that level, you will realize that there is no institutional or political force that makes known there are alternate technologies available for developing or developed countries. Thus ITDG, as well as other appropriate technology organizations in the world, is trying to fill this knowledge gap.

Practical Alternatives

We have a good deal of experience in this. For example, we have been asked by various African countries if we can provide them with detailed information about the technology and the costs of small-scale plants for sugar processing, glass manufacturing, brickworks, and cement substitutes. The only one that we were able to answer effectively was with brickworks, because we actually had a hand in making and running brickworks in Ghana. In the other cases the knowledge is fragmentary or dispersed and far away; you can in fact only get information about small-scale sugar plants in India. By the time the correspondence has gone to and fro, the big sugar companies are there breathing down the neck of the administrator and saying, "Sign here. A turnkey project. The whole thing will be operating within a year." The administrator faced with the need to make some decision quickly will opt for the high technology because that is given him on a plate. No one is there saying: "There are plenty of alternatives and here they are,

RCN: 0-87857-188-4   Book   Diskette Record Number = 47

008 ―― 770729s1977   paua   b   00100 eng
010 ―― ‡a   77010596 //r842
020 ―― ‡a08785718984 ‡c$6.95
050 0― ‡aT49.5‡b.I58
082 ―― ‡a338.91/172/4
245 00 ‡aIntroduction to appropriate technology :‡btoward a
        simpler life-style /‡cedited by R. J. Congdon ; ill. by
        Roy H. McCullagh.
260 0― ‡aEmmaus, Pa. :‡bRodale Press,‡cc1977.
300 ―― ‡axviii, 205 p. :‡bill. ;‡c28 cm.
500 ―― ‡aincludes index.
504 ―― ‡aBibliography :‡p. 200-202.
650 ―0 ‡aAppropriate technology.
650 ―0 ‡aLife style.
700 10 ‡aCongdon, R. J.
55   ―― ‡q0-87857-188-4

and these are the costs, and here is the operating experience of two or three countries where they have been tried out." Or "Here is the team of people who will come and put it up and train your people in running it." None of that happens, and so country after country decides in favor of large-scale projects, although they know, as in the case of Tanzania, that that is precisely the wrong thing to do.

Levels of Need

The knowledge gap is a very wide one and it ranges right down to the level that is outside the market. This sector below the market level is where knowledge of how to do things "without any recourse to the market" is most needed. I think this sector is important and is one of the areas that economists have totally neglected. Over a large area of the developing world the problem people have is not how to spend their money to best advantage. On the other hand, the big problem in rich countries is how to get the best bargain; we are a bargain-hunting society because we have income, and our problem, and a great deal of the problems of economics, is how to distribute that income to greatest advantage. But for the man who is unemployed and has no income, his problem is not how to distribute his money to greatest advantage, because he has not got any. His problem is how to turn his labor into something useful and that is quite a different problem and not one, I suggest, that can be studied within the context of economics as we now know it. So there is first of all the need to cater to nonmarket economy by providing a technology which enables people to turn their labor into something useful with the minimum possible of imported materials.

Above this, of course, there is the next level, which we might call the community level, where a group of people can get together, to try to start their own small production unit. And above this again is the market town or regional center level, which would utilize what is now recognized as small industry.

Now all of this is below the level of high technology as we at present peddle it to developing countries. So the gap which we are trying to fill is a very wide one. This also accounts for a lot of confusion about what is meant by intermediate technology or appropriate technology. Thus some people have in their minds only the simple hand-tools, the do-it-yourself type of tool; other people think only of small industry. And of course we are concerning ourselves with both, and the one in the middle as well, because these constitute the gap. We try to formalize this by saying there is no such thing as *the* or *an* intermediate or appropriate technology; there is a range of them. And the great thing is to provide information over that range and give people a choice. So at the first level the capital cost might be on the order of $20 to $40 per workplace, $40 to $400 at the next level, and then from $400 to $800 or $1,000.

Identifying the Gaps

The problem of identifying these gaps is of course a very real one, because you need to begin with some knowledge of what exists and a fair knowledge, at least, of what could be done. Again I think it is very easy to fall into the trap that the official aid-agencies have deliberately fallen into and this is to say, "We cannot do anything until we have received requests from developing countries." When we started the group the Minister said, "Just show us the demand for this simple level of technology and we will respond immediately." But the fallacy lies in the first part of the statement: "Show us the demand," because the only thing that the

This sketch shows how a catchment tank works. When rain falls, the water runs off the sloping catchment apron. It collects on top of the sand, which is piled a foot or so deep on top of the domes. The water soaks through the **sand, and trickles down into the cavities under the domes. To get water out, a bucket is lowered into the open well. Or a hand pump might be used. Sand filters the water both as it enters the tank, and again as it enters the well.**

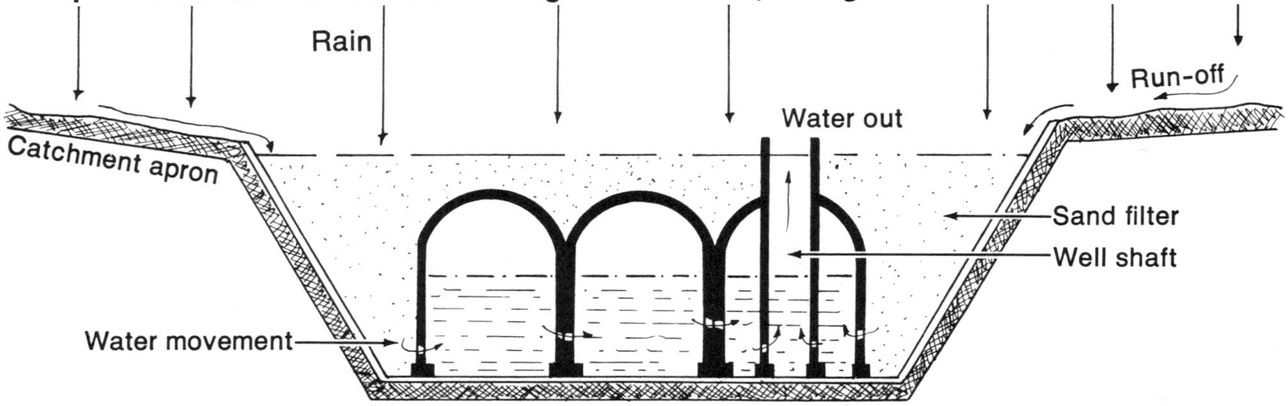

Rain

Run-off

Water out

Catchment apron

Sand filter

Well shaft

Water movement

Figure 1.1. Rainwater catchment tanks.

developing countries have knowledge of is the highest level of technology. Now since they do not know that there are alternatives, how can they be expected to ask for them? And this is really one of the first and most long-lived disputes that we have had with the official aid-agencies. If a man in central Africa, India, or Latin America wants an animal-drawn harrow, he had first of all to invent it in his own mind before he can ask for it. And if he can invent it, he would probably not be asking in any case. So the first task is to make known that there are alternatives and then a demand might be forthcoming. But then of course, one encounters the familiar argument used by the official aid-agencies, "We cannot do anything without a specific demand or request, and that request must come from the government. We only deal on a government-to-government basis and it must be a demand which is generated within the country itself."

Thus, there is a real task in making known that information exists, and such a process of mobilizing information cannot wait for specific demands to arise from developing countries. Fortunately, there is a very wide range of activities which many people throughout the world are trying to perform in one way or another and these activities relate broadly to the activities of growing and consuming food, clothing, shelter, health, culture, and certain basic community services. If you take these subjects alone and spell them out in terms of the manufacturing activities they imply, you get a very respectable range of industries for which information is needed all over the world.

Assembling and Disseminating Information

Now, until such information is made available, people will not be able to take hold of it and adapt it to their own circumstances. What the rich countries *cannot do* is to adapt technology to suit precisely the needs of particular areas, but *they can* at least make known that these alternatives exist and give some assistance in the process of adapting them to meet particular needs in particular areas.

There are different ways, of course, of trying to get this knowledge together. We have concentrated on bringing together groups for this purpose. We are also trying to bring in the universities, technical colleges, schools of art, design centers, professional associations, and quite recently, I am happy to say, a number of government research institutions. A number of such institutions which were supposed to be dealing with developing countries have now shown signs of beginning to collaborate with us in what we are trying to do.

Finally there is the dissemination of the information and this can take a number of forms. If the information is not published, nothing can happen at all because until the information is published, it does not really exist. So how do you publish it and to whom do you address these publications? We are trying to do this in various ways, particularly through a journal, *Appropriate Technology*. We are trying to use it as a vehicle for bringing together people who are working in this field. But there are much more effective ways of disseminating information and one is actual field demonstration. I do not mean carrying out enormous programs of development, that is the job of governments, United Nations agencies, and some of the very big agencies dealing with aid and relief. I mean demonstrating that these technologies actually work in practice; this means forming links with countries overseas.

Overseas Links

Each country has its own ways of making links. The Netherlands has obvious historical links with Indonesia, and has a particularly great advantage in dealing with most other countries, in that it is not

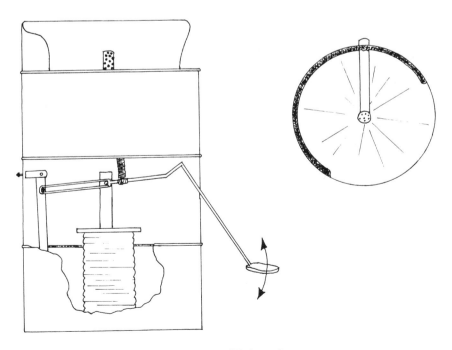

Figure 1.2. Oil drum forge.

seen as an ex-colonial power, for example in India, or Ceylon, or Latin America. Wherever these links can be formed they should be, in order to carry out demonstrations of the utility of particular technologies. They can be formed in a whole variety of ways. They may be in universities, they may be with interested government departments, they may be with voluntary agencies or with missions. In fact I would go so far as to say that the actual publication of information should, wherever possible, follow field testing. One should not publish information that has not been field-tested somewhere, not necessarily by yourselves but by somebody.

Overseas links, of course, are also essential for identifying and understanding local needs, attitudes, and conditions. In the case of agricultural equipment, for instance, virtually no work at all has been done on the identification of needs in rural areas for equipment, not even for tractors. We were astonished to find, when we started our work on agricultural equipment, that in spite of something like 20 years' work by British universities on agriculture in developing countries, nobody had ever gone into a rural community and said, "What are your problems?" We had always gone in and said, "We know what the answers are; the answer consists of a tractor. Please sign here." When they call it aid, I wonder who aids whom. Certainly a lot of aid has gone to tractor manufacturers in Britain.

Surveying Local Needs

The pattern of labor used in a typical rural community may be represented by the graph in figure 1.3. Although grossly oversimplified it shows the demand for labor throughout the agricultural year. Frequently, of course, there may be double-cropping which would bring two labor peaks close together; the second task may not be started until the first is completed, and naturally weather and the seasons will not wait for man to sort out his labor bottlenecks.

The problem of these communities really is to find ways of breaking these bottlenecks on increased production. At the points A, B, and C the whole community is involved, and with the existing level of technology they cannot produce more than they are doing. During the periods D and E they are grossly underemployed and this is the period when new activities need to be introduced into the rural area. They may not be farming activities but they will obviously be linked with farming. Now in order to discover what these bottlenecks are and precisely what they consist of, you will have to go into a rural area and examine it in great detail for at least a year, preferably two or three years, so that you cover a couple of good seasons and bad seasons. Not until you have done that can you have a picture of what their problems are. And our work so far has very clearly brought out this sort of pattern and the need at peak labor times to introduce improved mechanical equipment. This must be cheap enough to stand idle for long periods of the year, because it is only going to be used for a short time, and simple enough for people to make it during the slack period. If they have to import equipment, such as tractors, the development process is far more restricted. Tractors of course are indiscriminate; they put people out of work not only at A, B, and C but also at D and E.

Yet none of this work has been done for agriculture and it is likely that in almost every other field you consider a good deal of this kind of work would have to be done first. The alternative is to go in with our ready-made solutions. For instance, the classic argument about the tractor is that the ground is so hard after the dry season that only a tractor will break it

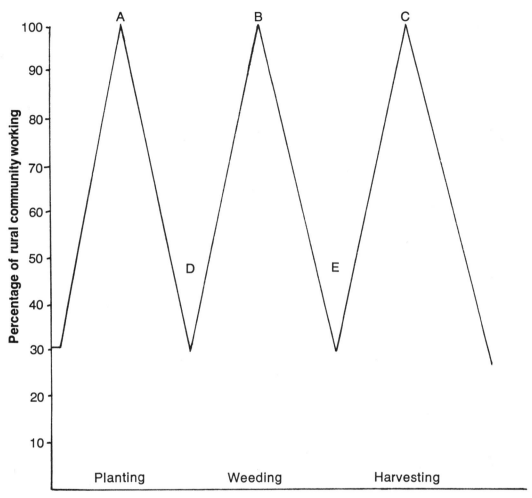

A, B, and C are the barriers to increased production.
D and E are the areas of un- and underemployment which
require low-cost and small-scale local manufacturing.

Figure 1.3

quickly enough to allow planting, the timing of which is absolutely critical to within a few days. The obvious answer to this is to change slightly the pattern of agriculture so as to avoid the need for a tractor. Our agricultural people, who have worked in East Africa for a long time, tell us that this has been done and is being done on increasing scale. You do not, in fact, turn under or clear away the rubbish lying on the ground immediately after the harvest. The answer is to run duck's-foot tines through the soil, just breaking up the first few inches of the soil and leaving all the rubbish on top. You leave it like that till the rains come, which then don't run off the soil. The rubbish becomes part of the humus and the ground is already broken up. There is no need for tractors provided one approaches the system intelligently. In addition, the equipment is very simple and can be made locally. So in almost every case the arguments that are used for high technology do not stand up to any sort of investigation.

And yet this sort of approach has only just begun. I think our group was the first to adopt it, but it is now being taken up by other people and one hopes it is going to be done on a much larger scale. The chances are that one needs to do the same sort of careful investigation for other activities in order to arrive at precisely what will suit the community. And there is every reason to suspect that similar questions are going to be raised more frequently in industrialized countries.

In order to arrive at precisely what will suit the community you have to undertake the type of investigation we have just been considering. But in order to make available the range of choices which will solve those problems, you can begin work without waiting for the demand. When our people went out to Zambia for this agricultural investigation, they took with them 40 designs of equipment which looked applicable to the local problems. These designs were drawn from 50 different agri-

cultural research centers throughout the world. Not all the designs were of any use but some were, at least by giving people ideas of how the equipment might be adapted or developed further. Adaptation is important. There are plows that have been developed for use on the east coast of India, where the bullocks weigh about a ton and stand six feet at the shoulder. The bullocks in drier parts of Zambia more resemble greyhounds. So the sort of equipment they can draw has to be quite different from the sort that big animals can draw. Then you have to take into account soil conditions and climate, social conditions and community obligations in adapting equipment precisely to local needs. And you need people who speak the language and understand the local community. But more generally one can

Photo 1.10. Concrete boatbuilding.

Photo 1.11. Completed concrete boat.

identify dozens of technologies on which people could be working now and for which I can assure you there would be an enormous demand in developing countries if only it were known that there were such choices.

Local Centers

We do not know nearly enough about the dissemination of such information. We do know, however, that publishing it is one thing, but the only effective way of making alternatives available on a large scale is to set up in developing countries themselves something like the Technology Consultancy Center which has been established in Ghana. This is a receiving point for information that can come from a number of sources in the rich countries, and is an essential link in the process. Local adaptation can in fact only really effectively be done on a large scale by local people. Certain trials and demonstrations can be done from outside, but the widespread introduction of changes can only come about from inside the country. So over the past few years we have been trying to concentrate on setting up equivalent organizations overseas, with a certain amount of success. They have varied in type. In Kumasi in Ghana, there is a university which has become a center of technical development for its surrounding area. Those of you who know the universities in developing countries will

13

recognize the extent of the breakthrough when I say that Kumasi is actually building prototype small workshops on its campus—the holy campus itself. It was originally designed to look as much like a campus in the United Kingdom as possible, but is now desecrated by actually having things built on it, such as small workshops, looms, and brickworks. This is a major break with tradition and we are very proud of it. It might spread.

In other cases governments have taken up the responsibility. Pakistan has set up what promises to be a very large, perhaps too large, center for the development of appropriate technologies. Tanzania already has something similar in the Small Industries Development Organization, and I gather that very soon the Eindhoven Technische Hogeschool will be having links with the Development Technology Center at the Bandung Institute of Technology. These overseas points are absolutely essential, because without them to whom are you addressing yourself? You can address yourself piecemeal to administrators and fieldworkers, but in order to work within developing countries on mobilizing their own resources for technological development, this sort of center is, I think, absolutely essential.

New Developments

I want now to mention some possibilities for new work that needs to be done. I mentioned earlier the basic needs: food, clothing, shelter, basic community services, and so on. On the basis of our experience I think if anyone said to us, "You cannot make it small," we would now reply, "Prove it!" Although I do not want to exaggerate how much has actually been done, what experience we have had indicates that anything can be made small and still be efficient. I was delighted

recently by our water engineer who has been working on the hydraulic ram. You can put this in any slow-moving body of water and it uses the power of the water itself gradually to pump water up to a distance of maybe 40 meters. Very expensive; the cheapest one made in Britain costs $400 and by the time it reaches the developing country you can be quite sure it costs $800. Simon Watt has now produced one costing less than $20 which can be made out of galvanized iron piping.

This is the sort of information we want to assemble. We should try to produce information on a technological spectrum, to have one of $8, one of $20, and one of $100. That is what would fit into my range. Make the range and then produce detailed practical information and make it freely available to developing countries. As another example we have our agricultural engineer who made a metal-bending machine. The nearest thing to it made in Britain cost $1,400 and he produced one for $14. Using principles that are perfectly well known, wheels up to 1.2 meters in diameter can now be made by any blacksmith with the assistance of this simple equipment.

Another area of considerable interest is the manufacture of cement. We are interested in two things: one is the scaling down of big cement plants and the other is making cement substitutes. The scientists employed by the cement industry are very angry about this because they say that outsiders have no business to meddle in the making of cement. From what little I know about it, the process is so simple that I don't know what they feel so worried about. But of course it is not really quite so simple. A remarkable man, M. K. Garg in India has done a lot of work on this subject and has produced a small-scale cement plant that works, but does not quite produce the right quality of cement. He has problems of temperature control and so on.

His work needs to be continued until a small-scale unit is perfected that is perfectly possible. There may be such cases, in which case the thing is to look for an alternative to cement. We are trying to pursue both approaches at the moment. There are alternative cement substitutes such as mortar mixes, which are perfectly well known—they need a bit more experimental work done on them—which can be used on buildings up to two or three stories high. Just as good as cement, but such is the power of the idea that you cannot do anything until you have got a cement plant, that developing countries all over the world have sat crippled and helpless because the big cement plant has not come into production.

Limits to Large Scale

The day of the large-scale unit is probably coming to an end because of energy transport costs. We have developed these huge technologies on the basis of cheap transport. I cannot otherwise explain the phenomenon that you can observe anywhere in Britain, and I am quite sure in Holland too; you stand on a motorway and you will see lorries from London to Glasgow carrying biscuits and lorries from Glasgow to London carrying biscuits (or any other product you like to think of). And the only conclusion you can come to is that there is something in the nature, the character, of biscuits that requires them to travel 500 miles before they are properly mature. There is no other logical explanation for the phenomenon. And such a system is only possible because of large-scale production, the curious form of competitive situation that we have developed, with transport and energy costs that were minimal. Until recently the costs of energy in manufacturing industry all over Europe was an almost insignificant proportion of the total cost. For an average

manufacturing industry in Britain it was about four percent, so they did not mind at all if it went up by 10 percent or not. Now, of course, energy costs are not insignificant nor are transport costs. As a result one can see that all sorts of ideas about large-scale industry and about agriculture are going to change very rapidly.

Perhaps I should end off with the cheerful thought that what we in the West have really done is to create a society which is based totally on the assumption of cheap and almost nil-cost energy. That is what gave rise to the growth of cities. Cities can only become very large if it takes only a very small proportion of the total population, food producers, to feed non-food producers. If it takes 80 percent of your population to feed 20 percent, then only 20 percent can live in the cities. But if you introduce a system of production, as we have done, based on cheap oil, which has a very high productivity per man (which is not the same as productivity per hectare), you can then have four percent of your population feeding 96 percent, and those 96 percent can live in cities. But the system of agriculture on which this process is based does not seem to me to be a permanent one, because it depends entirely on oil for its operation. Some work recently done in the United States shows that for several food crops it takes roughly one unit of oil energy to produce one unit of food energy. Now whatever else that is, it is not a permanent system of agriculture. But it is what we were on the threshold of exporting to developing countries on a very big scale under the name of the "Green Revolution" and other similar schemes.

Fortunately those intelligent men of the Organization of Petroleum Exporting Countries have saved us from such a course, and they may very well have saved us in other ways as well, unless of course they drive us to the even greater lunacy of atomic energy. We may be hooked on energy-intensive technologies, but at least developing countries have an option. There is no need for them to follow precisely the same path, and if there is nothing else we can be doing, at least we can be saying, "You may go in our direction; you may want to do as we have done, but at least we can offer you options. We can offer you a choice." And I think this is basically the question we are trying to answer—How do you actually set about the job of offering choices which are not theories, which are tested out and are practical as far as it is possible to make them?

But there is another dimension to appropriate technology that many of us are beginning to perceive. In the best sense appropriate technology offers alternatives to individuals and communities. Nonetheless, in the early years of this "movement," the emphasis was on deriving alternatives for developing nations. Recently, however, particularly with the advent of a National Center for Appropriate Technology in America, there is a budding awareness that even highly industrialized communities might well need some alternative technologies, that even Americans might be looking for alternatives to high-technology living.

Chapter Two

The Social Context for Choosing Water Technologies

by Simon B. Watt

At first sight it looks as though my chapter should simply be concerned with civil engineering from the point of view of water supply. But I hope you will appreciate by the end that I really believe that the actual technique or skill of engineering is one of the least important considerations. I want to concentrate much more on the attitudes and motivations of the engineer trying to implement an appropriate technology—in this case, a technology for rural water supply. I also want to examine ways in which the broad objectives of the research or implementation can be better understood by the engineers involved in it. This is particularly important in relation to the ways in which preconceived ideas of standards or "efficiency" can limit the extent to which the objectives can be met.

Know Yourself

I am not going to give a systematic description of appropriate techniques, nor an evaluation of the available technology. I think this would be misleading. Any competent engineer, with the right experience can locate and utilize commercial hardware, or if this is not appropriate, he can design and build suitable systems and equipment. Rather, I want to concentrate on his "education," how he is trained to conceive the problem, and how far he is able to look beyond conventional solutions.

The message of the children's story about the small boy who was not afraid to point out that the emperor wore no clothes is important for us if we are to approach the problems of development with a truly open mind. I have changed the details but the content is much the same.

The Small Boy and the Emperor with No Clothes

1. The best way to help the developing countries is not by aid, but by changing the terms of trade in their favor.
2. Developing countries would do better if they went to Chairman Mao's China before seeking help in Europe.
3. The professionals have become colonials in the sense that they have taken possession of the knowledge of technology—a knowledge that all people should possess to be able to change their own lives.
4. We should put our own backyards in order before we look at other people's.
5. Mankind is a thermodynamic species like all other animals on this earth.
6. The purpose of economics is not to maximize the flow of resources in through our mouths and out through our waste disposal systems.
7. Small boys are not afraid of being fired.

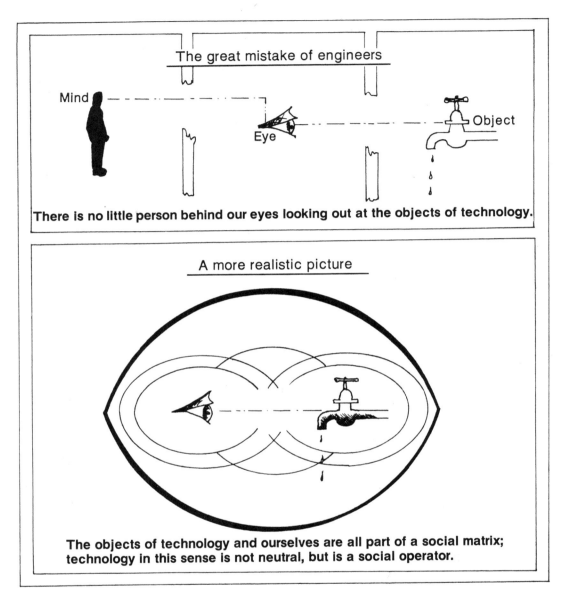

The great mistake of engineers

Mind

Eye

Object

There is no little person behind our eyes looking out at the objects of technology.

A more realistic picture

The objects of technology and ourselves are all part of a social matrix; technology in this sense is not neutral, but is a social operator.

Figure 2.1.

What this little parable shows is that the first requirement in the development process is that you should discover who you really are yourself, and what are the assumptions and prejudices that influence your position and subsequent work. When I am asked what is the most important qualification and training a person needs before he is fit to work overseas, the only answer I can ever give is that he should simply be a human being. But the conventional engineering approach views technical problems much more in the fashion of the first sketch in figure 2.1. Engineers tend to see themselves as masterminds having an "objective" picture of the world and its problems. The second sketch is in my view more realistic. It depicts the objects of technology and ourselves as being all part of one social matrix. Technology in this sense is not neutral; it is a social operator.

The analysis of the effects of technology on society, and the reverse effects, has only just begun. Little change has yet come to the way engineers are trained. We do not yet see an interaction between society and technology. Nor do we yet appreciate the range of alternatives that are open to us if we are prepared to change our life-styles.

Now there is, of course, a range of technologies from which it is possible, though it is not often done, to choose the most appropriate in any situation. Figure 2.2 compares the attributes of high-perfor-

mance and low-performance technology. The difference between the two is the quick fall-off in the benefits derived from high technology when its operational parameters are not favorable. This is only another way of saying that high technology is generally conceived in, and designed for, a particular society and a particular set of circumstances. One should, then, not be surprised if it fails to produce the intended results in different circumstances.

Figure 2.3 takes the notion a step further and portrays a spectrum of technology. And in evaluating the benefits of a technology—performance, output, efficiency—one should be continually mindful of the second law of thermodynamics; one cannot get something for nothing. Of

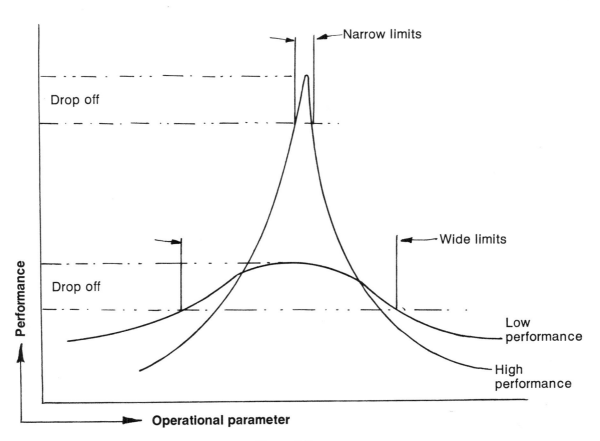

Figure 2.2.

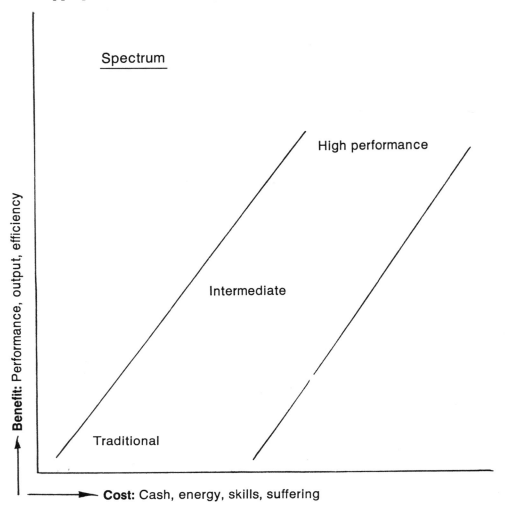

These are separate relationships. Combining them is very difficult. We have no quality-of-living index.

Figure 2.3.

course I should not give the impression that all technological "progress" is universally acclaimed. People generally are a little conservative and afraid in their reaction to technology—"Are we actually going to gain anything by this development, how will it affect our level of welfare?" Or, if people really talked in the language of the economist, they might say "What degree of disbenefit do we expect?"

To be able to evaluate these costs and benefits in quantitative terms is, of course, extremely difficult. And indeed the objective of this discussion is only to show some of the relationships that exist between the various factors. Such an analysis is further complicated by the fact that not every factor is capable of being measured. In spite of this the things that can be measured are not always accepted, or acceptable. For example, some transport studies that have been carried out in Britain show that 80 percent of the costs spent on improving transport benefit only 15 percent of the country's population. This does not seem very equitable, and one hopes that such a result was not the original intention. There are various devices that can be used to overcome some of the problems of comparing dissimilar things, such as the economists' technique of shadow pricing. But techniques like this do not really seem to face up to the reality of the problems they are being used on. And the consequences of this way of thinking simply in numerical terms can be more or less disastrous. There has been discussion recently in Britain over the policy of "shooting lame ducks," which suggests failing industries should receive no government support and should simply be allowed to go bankrupt, on the basis that if they are inefficient, they do not deserve to survive. Opponents of this policy, rightly in my view, point to the costs of unemployment which result from such a policy. To take another example of this numerical neurosis, the United Nations has tried to devise an index by which the "quality of life" may be measured. The result of this exercise was to show that a life in the United States was "worth" approximately five times a life in India. Not unnaturally this led to something of a conflict, and the discussion was abruptly stopped.

Choosing Appropriately

It is now time to move on from these general considerations to see what impact they have on the developing world and to explore what possibilities they have for making us look at development in a different, and perhaps more sympathetic, way. One example that I have been involved in is the scheme supported by the World Bank for pumps to be used in Bangladesh to supply water in the dry season through an average head of three meters. The various alternatives that were considered, together with their shadow prices, are shown in figure 2.4. A shadow price simply means that $X in country A can be considered as equivalent to $Y in country B, and this can be used to differentiate between "real" and local costs. The World Bank's choice was for the second of the alternatives shown in the figure, while I suggested that a cheaper solution should be sought. Actual experience of the implemented scheme has shown that only about one in every one hundred installations has remained in use. This is because of various shortages in spares, corruption, lack of maintenance, and so on. In this scheme a lot of the money of the original investment has effectively been lost, but rather than become introspective over past failures, we should ask what we can do to prevent a similar type of situation arising again. At the very least in this example we should be studying how low-skill maintenance in high-performance technology will inevitably lead to a fast reduction in output.

Figure 2.5 shows the sort of solution that I would favor. This particular solution illustrates a simply dug well and a cheap and simple pump—perhaps a chain pump powered by humans. Figure 2.6 illustrates a similar sort of approach for a well screen. It was originated or "invented" by a farmer, but the idea of such a primitive bamboo screen was rejected, almost out of hand, by the "experts." Nonetheless this innovation works and has achieved widespread acceptance. So much so that, unlikely as it seems, it is a solution being recommended by the United Kingdom's Atomic Energy Authority whose government scientists now find their underemployed skills being directed towards the Third World.

Such a fundamentalist approach should give us ideas about the way our own development could proceed in the industrialized countries. It is not a foregone conclusion that a technical possibility must lead to a particular form of implementation. After all the Chinese knew all about gunpowder long before it was used

Comparison of cost of three tubewells of varying specifications in market and shadow prices in Rupees; shadow prices (in the United States). Rs. 9.50 = $1. Dollar values in ( ).

drilling — screen — pump — engine —	jet/percussion brass centrifugal low-speed diesel		contractor/power fiberglass turbine high-speed diesel		contractor/power fiberglass turbine electric (including generation/trans- mission cost)	
move in and out	300	(150)	4,500	(4,500)	4,500	(4,500)
drilling cost 160 feet 180 feet	1,440	(720)	7,600	(13,300)	7,600	(13,300)
pumphousing 10" dia. 40 feet			4,800	(9,600)	4,800	(9,600)
blind pipe 8" dia. 20 feet	800	(1,600)				
screen 80 feet 140 feet	7,000	(11,900)	4,280	(13,482)	4,280	(13,482)
bail plug and reducer	500	(1,000)	500	(1,000)	500	(1,000)
gravel pack 80 feet 140 feet	2,400	(2,400)	1,200	(1,200)	1,200	(1,200)
install well hardware 160 feet 180 feet	2,160	(1,080)	1,920	(1,920)	1,920	(1,920)
develop and test well	1,500	(750)	1,500	(1,500)	1,500	(1,500)
pump	750	(1,125)	5,500	(11,000)	5,500	(11,000)
engine 820 h.p.	4,500	(6,750)	6,000	(12,000)	125,000	(215,000)
right-angle gear drive			1,500	(3,000)	1,500	(3,000)
pumphouse	3,500	(3,500)	4,500	(4,500)	3,500	(3,500)
install pump and engine	750	(375)	750	(750)	750	(750)
consultants			5,000	(10,000)	10,000	(20,000)
field distribution system	3,500	(1,750)	3,500	(1,750)	5,000	(5,000)
contingency	2,560	(2,560)	4,955	(4,955)	17,255	(29,975)
TOTAL Rupees	31,660	(35,660)	58,005	(94,457)	194,805	(334,727)
or Dollars	3,300	(3,700)	6,100	(9,900)	20,500	(3,500)

Figure 2.4.

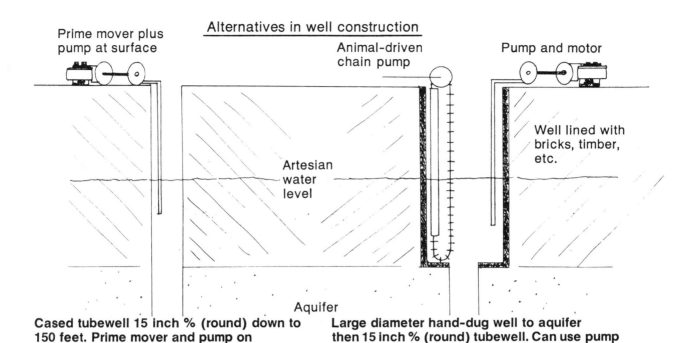

Alternatives in well construction

Prime mover plus pump at surface

Animal-driven chain pump

Pump and motor

Well lined with bricks, timber, etc.

Artesian water level

Aquifer

Cased tubewell 15 inch % (round) down to 150 feet. Prime mover and pump on surface.

Large diameter hand-dug well to aquifer then 15 inch % (round) tubewell. Can use pump and prime mover or simple local equipment powered by animals, etc.

Figure 2.5. Alternatives in well construction.

in warfare. And other countries can, and of course do, evaluate their costs and benefits in different ways. The Chinese, for example, don't shoot lame ducks; instead they "walk on two legs" which represents the policy of encouraging development wherever it can take place, whether in the urban or the rural areas. In other words China encourages broad-based, and therefore decentralized, growth while Europe tends, on political grounds, to cut off the weaker parts.

Problem Definition

Another consequence of studying the hidden implications of the problem is that something that looks superficially like a simple and conventional engineering problem may turn out to be far more complicated. Suppose, for example, that we are trying to develop a cheaper pump maximizing the use of local materials. This may very well lead to the partial problem of developing a suitable flywheel. Generally in Europe we would approach the design of a flywheel in terms of materials like iron or steel which have sufficient strength to enable large-sized wheels to be built. But if we translate our design into other materials which have less weight and less strength, the design will become more expensive because the size and the weight will increase without the strength being increased. This means that

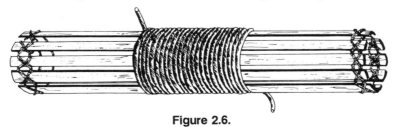

Figure 2.6.

23

Relative Costs of Brewing (per 15p pint)		
	Home Brew	**Charles Wells**
Raw materials	0.32	0.80
Brewing costs (labor, transport, overhead, equipment)	2.50 write off equipment at 1 pint per day for 1 year and 6 minutes/pint to make beer	2.90
Brewer's profit		2.10
Licensee's profit		4.30
Tax: Vat		1.50
Excise duty		3.40
	Cost — 2.8 pence	— 15.0 pence

We should ask what are the effects of these methods of production on: local employment and industry, flow of materials, quality and quantity of beer, cost to imbiber, cost to community, loss to treasury, effect on social habits, etc.

Figure 2.7.

we would have to investigate new constructions for flywheels. And to do this it would be necessary to understand the real purpose of a flywheel. So what looked like a simple redesign of a pump has resulted in our exploring the range of possible energy accumulators.

Perhaps the rather down-to-earth example in figure 2.7 will help to clarify this general point about the need to understand not only the objectives of the exercise but also the implications of the chosen technology. To arrive at a comparable state of inebriation the cost of home-brewed beer is approximately six cents while commercial beer would cost thirty cents. But one cannot simply make the comparison in terms of the same "benefits" for differing "costs." We should also investigate the implications of the chosen method of production, and the effects it will have on such factors as local employment, industrialization, the flow of materials, transport, and many other such things. One has to be able to have an overview of the whole spectrum of technology that is available for solving the problem and then to make a thorough cost/benefit analysis before one is in a position to be able to select the proper, or most "appropriate," technology.

Standards

Some people maintain that an "appropriate" solution would be incompatible with the idea of a piped water supply. But if one studies these criteria, one sees that this is not in fact the case. The appropriateness of different technologies will vary with each situation. A range of some of the alternatives is shown in figure 2.8. This displays relative costs for various levels of technique. There is a tendency for attention to be given to the most sophisticated systems. For instance the World Health Organization demands that certain standards of water supply should be met in any new scheme. Clearly to ensure that all those in need of improvement in their water supply and sanitation have a system which meets these high standards would be prohibitively expensive. At that level of technology, and cost, it would be impossible to connect all houses to centralized water supply, drainage, and sewage systems. And there is no guarantee that doing this would produce the benefits expected. As an illustration I feel that the social function that accrues to a village well needs to be re-investigated. In real life many village women prefer to walk three kilometers to a traditional water supply so as to be able to continue their usual pattern of meeting and chatting to other women rather than use a newly built village well.

The important lesson to learn is that there is no uniquely applicable appropriate technology, no one solution that will apply in all circumstances. Appropriate technology implies a step-wise development, and investment will have to be made over a wide range of levels. In this sense the choice is a matter of the allocation of scarce resources. It is thus much more a political than a technological choice.

Criteria for an Appropriate Technology

The water supply and sanitation technology chosen (or developed by research) for rural areas and small communities should:

1. Facilitate significant improvement in quality and quantity of service without necessarily seeking to obtain the near-perfect
2. Be low in cost; as low as possible without jeopardizing the effectiveness of the improvements sought
3. Facilitate operation and maintenance by local populations and users without demanding a high level of technical skill
4. Make as much use as possible of locally available materials, and rely as little as possible on imported supplies, spare parts, and equipment
5. Make use of locally available labor, including unskilled labor and not try to replace labor with capital equipment unless it is clearly imperative to do so, e.g., when local labor is scarce
6. Encourage the growth of local manufacture to supply the need for equipment and parts under the leadership of local entrepreneurs
7. Be compatible with local and user values, attitudes, and preferences
8. Provide opportunity for incremental adoption and step-by-step improvement
9. Have a capacity for producing a contagion effect and so diffusing to other communities and individuals
10. Facilitate community involvement and participation.

Stages of Improvement

Type of Supply	Technique	Quality	Management	Cost/Capita $ Low Median High		
No improvement	None	No protection	Individual	0	0	0
Individual improvement	Spring, well, pond, rain collection	Curb major pollution	Individual	1	6	30
Group improvement	Spring, well, pond, rain collection, pipeline	Curb major pollution	Group untrained, maintenance unreliable	1	3	14
Rural pipeline	Piped to taps and standpipes	Moderate protection	Unskilled but regular operator	3	5	10
Municipal standpipe	Piped to taps and standpipes	Good protection by disinfection	Skilled trained operators	3	7	20
Single tap	Piped to individual households	Good protection by disinfection	Skilled trained operators	5	12	20
Multiple tap	Piped to households, multiple taps	Good protection by disinfection	Skilled trained operators	8	25	300

(After White, Bradley, White, 1972)

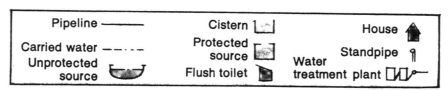

Pipeline ———	Cistern	House
Carried water —·—·—	Protected source	Standpipe
Unprotected source	Flush toilet	Water treatment plant

Figure 2.8.

Health

In the details of safe water supply shown in figure 2.9, we can see that in 1970, 88 percent of the population in developing countries had no access to safe water. The prediction for 1980 shows a superficially spectacular increase in the percentage expected to have such access, but virtually no improvement in the numbers without a safe supply. This is a result of population increasing faster than improved supply can keep pace. The sketches included in this figure show the relations of health benefit to water supply. We must consider about two liters per day as the minimum requirement, whereas in the West approximately sixty liters per head per day is used. This is considerably more than is actually needed, or at least we can say that the improvement in health benefit above a consumption of about 20 liters per day is negligible. In the face of the great disparity in consumption rates, and the increasing cost and difficulty of meeting the West's continually rising demand, it would be a major achievement to change from a policy of supply increase to one of demand decrease. A reduction of consumption would at least overcome the West's water-supply problems, even if an increase in consumption, so necessary for the developing world, would be much more difficult to achieve.

It is reasonable to suppose that health is a function at least of the volume and quality of the water available. If only small improvements were made in many existing supplies much current illness and disease could be controlled. We must, therefore, not look on water as being only for the purposes of cooking and cleaning in the house; it also has an important role in health care in the illness prevention sense.

Now the cost of water per liter per head of population increases sharply as one increases the proportion of time for which the supply is maintained. One approach to safe water supply, then, might be to admit that during certain periods of the year we could bear unpleasant conditions or a little bit of illness. In other words we might think of increasing the periods of health by increasing the periods in which good water is available.

It seems clear, then, that when one examines the question of the standards that one is going to adopt in new water-supply schemes, one must choose an improvement that will in reality enable the desired benefits to be attained, and that will allow the maximum number of people to enjoy these benefits. Unless one adopts an appropriate approach that can take these benefits to the rural areas, it is inevitable that the trends of urbanization will continue and we will end up with one-half of the population in overloaded cities and at least one-third of the population without any job or income.

Hygiene vs. Medicine

Figure 2.10 shows the mortality trends in Britain. For a good number of years the importance of hygiene in this respect has been well understood. Of course water is not the only requirement for good hygiene—education about the use of soap and the washing of hands, or the control of tuberculosis in milk are also significant. In fact, we have to introduce a whole system—water supply, user's facilities, disposal, soap, and so on. It achieves nothing to set up just one small part of the system as development aid.

But an examination of the system in the United Kingdom shows that the investment has now become involved in what might be called death prevention, rather than what should be called health improvement. In figure 2.11 we see the reduction in the mortality rate among

Program for rural water supply in 90 developing countries   1970/80

(population in millions)

Type of supply	1970		1980		Increase 1970/80	
	No.	%	No.	%	No.	%
Access to safe water	140	12	357	25	217	155
Without access to safe supply	1,026	88	1,081	75	55	5
Total population	1,166	100	1,438	100	272	23

Health benefits

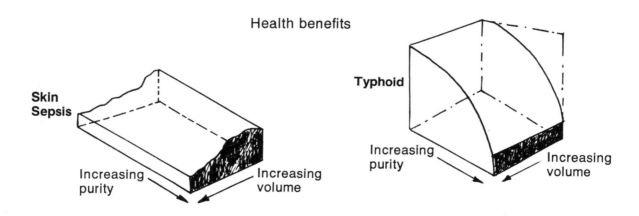

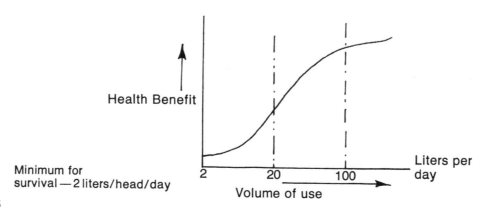

Minimum for survival — 2 liters/head/day

Figure 2.9.

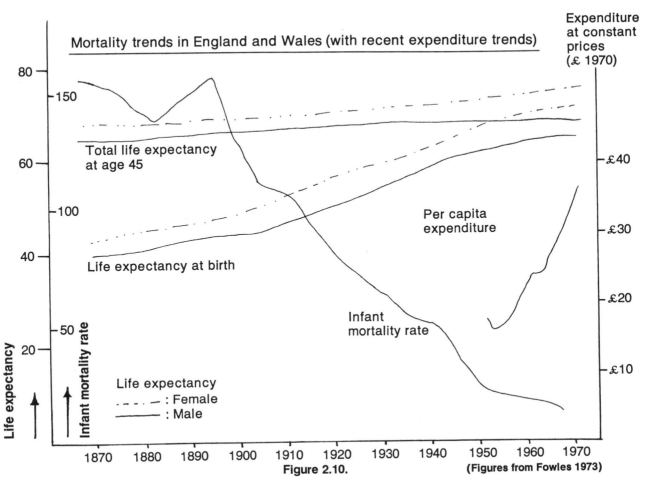

Mortality trends in England and Wales (with recent expenditure trends)

Expenditure at constant prices (£ 1970)

Total life expectancy at age 45

Per capita expenditure

Life expectancy at birth

Infant mortality rate

Life expectancy

Infant mortality rate

Life expectancy
— · — · — : Female
———— : Male

Figure 2.10. (Figures from Fowles 1973)

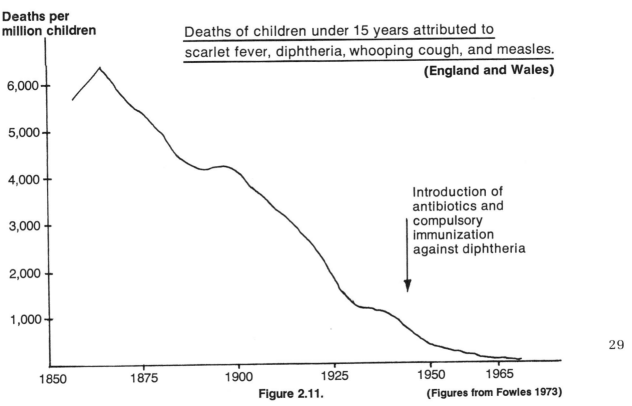

Deaths per million children

Deaths of children under 15 years attributed to scarlet fever, diphtheria, whooping cough, and measles.
(England and Wales)

Introduction of antibiotics and compulsory immunization against diphtheria

Figure 2.11. (Figures from Fowles 1973)

29

Preventive Measures for Water-Related Diseases	
Disease Category	Preventive Strategy
Water washed	Improve water quantity and accessibility Improve standards of hygiene
Water borne	Improve water quality Prevent use of contaminated source
Water based	Decrease water contact Control pest populations Improve water quality
Water related	Improve surface water management
Insect vector	Destroy breeding sites Decrease need to visit sites

Figure 2.12.

children under 15 years as a function of time. The introduction of antibiotics and compulsory immunization is indicated, but this has virtually no accelerating influence on the decline of the graph. The dramatic downturn over the period shown has largely to be accredited to improved water supply and sanitation which led to better hygiene. The effects of this are very large in comparison to the very small improvements resulting from immunization. This point leads to the question of whether, in development assistance, we should strive after health systems based on very expensive antibiotics and special intensive medical care or after better water supply and sanitation. Figure 2.12 shows some of the measures that are needed to prevent various water-related diseases. The technology required for these strategies is not a complex one, although their implementation needs widespread motivation and training.

My general conclusion is that in terms of health care one has to make a choice for investments aimed at activities which come within the systems approach and can make improvements in small steps. Such a choice will lead one to questions like the value of immunization compared to incremental improvements of water supply. And, of course, engineers should be aware that public health engineering is very closely related to preventive medicine.

Centralization

Three graphs are given in figure 2.13 that show economies of scale in central service plants, economies of density in dis-

I. Economies of scale in central service plant (UK)

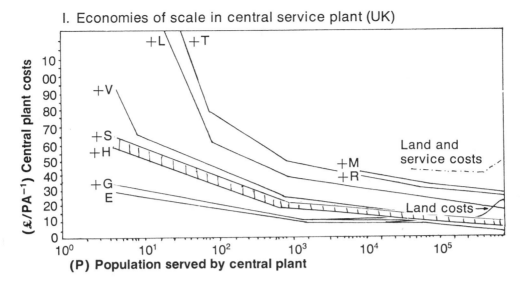

II. Economies of density in distribution systems

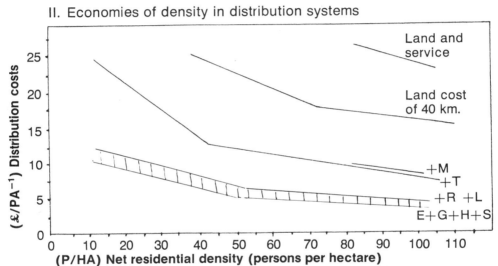

III. Economies of location in service transmission systems

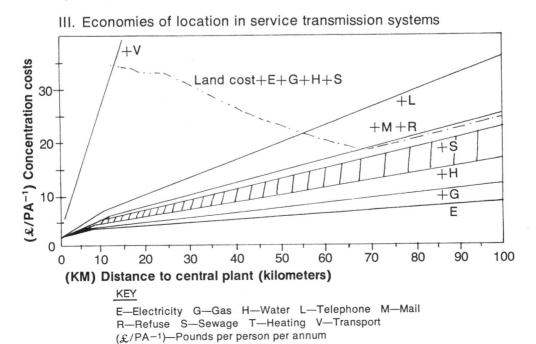

KEY

E—Electricity G—Gas H—Water L—Telephone M—Mail
R—Refuse S—Sewage T—Heating V—Transport
(£/PA⁻¹)—Pounds per person per annum
▨ —Water plus sanitation

Figure 2.13.

31

Incremental improvement of water supplies

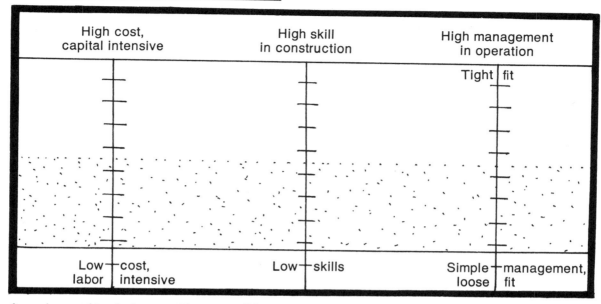

A package of techniques with compatible costs and operation skills is presented to the users.

Figure 2.14.

tribution systems, and economies of location in service transmission systems. They indicate that a large concentration of population is needed to make a cheap health service possible if it is to comply with Western standards. Indeed one might say that large concentrations of population are a consequence of the Western way of thinking, because if population is not evenly distributed nor concentrated, Western service systems will be very expensive and would probably be impossible.

Investment strategy should be based on a decision-making model which is not just a rural/urban model; it should extend much deeper. Current development aid policies can and do result in the situation that, if an improvement is made in the cities following Western methods and using Western money, then 80 percent of the money that flows into the country ends up by benefitting mainly the five percent of the population that lives in the cities. Many developing countries say that they want low-cost development that will be applicable in the rural areas, but they seek to

achieve this through Western aid or Western technology. And aid is a commercial commodity; in the United Kingdom the rate of return on our aid expenditure is 150 percent per year. One might rightly feel that the stated objectives of most development programs and the means chosen for achieving them are incompatible. Present planning techniques, however, make it difficult to do anything other than state that for objective "Y" we need an amount of money "X," and hope that next time we will have learned from our mistakes. We should also understand that the improvement of living standards can be a commercial asset for the poor. As the experience of slum improvement schemes near developing cities shows, the poor frequently prefer to sell out, to cash in the physical benefits they have received, in order to obtain a spendable income. So the improvement does not benefit the original target group but finally ends up in the hands of those already rich enough to buy their way into areas that have been improved through government assistance.

Small Steps

To return to the particular case of water supply schemes, figure 2.14 shows a schematic of incremental improvements in terms of capital, skill, and management. We should by now be agreed that we are looking for low-cost, labor-intensive systems requiring low skill and simple management. But it is so far only possible to indicate the range of possible choices in this theoretical way. One is not yet able to fill in the matrix and demonstrate that sufficient alternatives are available. Many people throughout the world recognize the problem, but few know how to tackle it. The failures of the old approach are being exposed in a general and analytical way but there will need to be a considerable amount of detailed and technical work done before satisfactory solutions can be implemented. The few people who are working on building up the body of knowledge, however, will benefit from a "multiplier" effect and one can already begin to see some movement and progress in the subject of water supply and sanitation for rural areas.

Chapter Three

Tools for Agriculture

by John Boyd

My objective here is to describe the work of the agricultural project of ITDG.

The way that the agricultural project has tried to make a practical contribution can be divided into six separate types of activities. First, we have published a buyers' guide to inform people in developing countries where they can buy simple pieces of farm machinery. Second, we answer postal inquiries, and act as a sort of clearinghouse for technical information. Third, we have published designs of machines so that people in developing countries may make them themselves. Fourth, we have published details of a system for carrying out field surveys to determine where the labor bottlenecks occur in a farming system, to identify where there is a peak labor demand so that one may be sure that one is introducing mechanization at the most appropriate point. Fifth, we have organized field projects, one in Zambia and one in Nigeria, which have put into practice some of the objectives and ideals of intermediate technology. These projects have involved developing simple farm machinery, encouraging farmers to use it, and training and stimulating local artisans to manufacture it. Sixth, and last, we have a large number of contacts with people who are very experienced in working in agriculture in developing countries, and we have on occasions supplied consultants for small projects.

Buyers' Guide

Having given a quick run-down of the agricultural projects activities, I shall explain in a little more detail what each one involves, starting with the buyers' guide for simple machinery. In the early days of the Intermediate Technology Development Group the main strategy was to publish information to tell people where they could buy pieces of simple equipment. The first major publication was called *Tools for Progress* which was followed by one on agricultural equipment, called *Tools for Agriculture*. This contains details of about 150 implements grouped in classes such as crop protection or cultivation. There is a short description of the characteristics of the tools in each class and an explanation of where they can be bought. Most of the entries are from Europe, in fact the vast majority from Britain, although there are one or two from Africa and India.

Technical Inquiries

The answering of technical inquiries was another activity that started early in the group's development. We have made contacts with something like 90 development institutions and individuals working in agriculture in developing countries. We have collected a lot of information from these people on simple farm equipment and have systematically searched journals for published information. We have managed to assemble a fair collection of photographs of implements, which has been most useful; in many cases people have sent them to us, in some cases our own staff has traveled overseas to collect them. Bob Mann, my predecessor, traveled through East and West Africa making two or three long trips a year collecting this information, making contacts, visiting people working in universities or research stations.

People constantly write to us for help in solving their problems. There is quite a wide range of subjects and of countries they come from. For instance, we have had an inquiry from somebody in Malawi, who wants to make a water-powered mill for grinding corn. Another inquirer from Tunisia wants information on small-scale storage, that is, silos for less than five tons of grain. Another person wrote from India wanting to know how he could make simple equipment for weighing cattle; he cannot afford to buy a machine to weigh them and has no metalworking equipment, only woodworking equipment, so he asked what possibilities are open to him. We had other inquiries about the "snail" which is a small tractor which has been recently developed partly in England and partly overseas. It uses the principle of pulling a plow with a winch, in the same way that steam-powered traction engines were used in the early days of powered agricultural equipment. One does not need to be an engineer to know the problem with small tractors generally is that if they have small wheels and they are light in weight, as soon as you try to pull anything the wheels will just spin. The "snail" can be anchored firmly to the ground and its winch applies the draft to the plow. In this way a very lightweight tractor can pull a heavy load. (See discussion of Rodale Winch in "Pedal Power.")

Designs

The logical development from dealing with individual technical inquiries was to progress to the publication of designs of simple farm machinery. The first problem, of course, is the acquisition of these designs. Some of them are collected on our tours, but more generally they come from exchanging information. We find that the most profitable way to get designs from experienced overseas centers is to give them two of our designs, and ask for two that they have developed in exchange.

In many other cases one can obtain the design rights of a machine which was formerly produced in Europe, but for which there is no longer a demand and therefore production has ceased. A particular example of this approach in which we were involved was for a bending machine to bend steel bars. This originated from an African inquirer who asked how to bend steel bars into circles to make a wheel for a bullock cart. We looked at the various pieces of equipment which might be available from Europe for this job, and the cheapest one that our project officer could find details about cost $1,400. But we managed to get the rights on a very simple design which could be constructed for $14, just one percent of the cost of the commercially manufactured machine. It is not as good in the sense that it does not work as fast and that it requires a lot of effort, but it cannot really go wrong. The design is simple and the cost of materials is very low. It can be made entirely from mild steel—no heat-treated alloy steels, for instance, are required. It is quite an achievement, I think, to produce a machine that is only one percent of the cost of the commercially available article. One or two other items we have developed in the workshops of the National College of Agricultural Engineering where the project is based, but most of the designs that we have published have come from the field projects in Zambia, Tanzania, and Nigeria.

So far, the policy has been only to publish designs which have already been built and tested in developing countries. It is still a matter of debate whether this is a good policy; should we only publish what has been proven to work in developing countries, or should we try and publish anything that we think might work, or that might give ideas to a local designer? We have really played for safety in publishing what has already been fairly thoroughly tested. The way in which we have published these designs has been somewhat unconventional. Only in a few cases have we used the normal sort of engineering blueprint on very large sheets. This is something which is usually only intelligible to a trained engineering technician.

What we did instead was to ask ourselves who might use our designs. And we thought there would be about four categories of people. It could be agricultural machinery factories, of which there are a few in developing countries; or it could be engineering departments in universities or in quite large research stations, which ought to be able to interpret manufacturing instructions in almost any form; or the designs might be used by much smaller research and advisory centers, perhaps government extension centers or missionary projects, which do not have any specialist engineers although they may

well have experienced "amateurs"; the last category of people who might use the designs are the very small workshops, village blacksmiths, and people who have graduated from being a village blacksmith, perhaps through learning a technique such as welding. Our experience has shown that this last category of people probably would not be able to interpret drawings in any form, let alone engineering drawings. Some of them are illiterate, at least in many cases they are not literate in Roman script, although they may be literate in a language written in Arabic or some other script. Most of them have very little idea of how to read and interpret a drawing and make it into a machine, although they can be very good at copying an existing machine. So our "target" has become the small research or advisory center which has no engineering staff—the place where there is just a "handyman," somebody with plenty of common sense but not a trained engineer.

The drawing in figure 3.1 can be reproduced on white standard-size paper. An advantage of this is that in any city, even in the middle of Africa, you can get

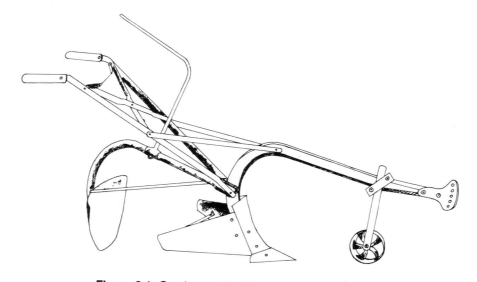

Figure 3.1. Ox-drawn, tie-ridger/weeder implement.

this photocopied and can print as many copies as you want. If one sends out a large engineering blueprint, the facilities for copying it are usually rather limited. Another feature is that there is no writing on these pages of drawings; there may be just a letter, or an arrow pointing somewhere. All the written instructions are on pages at the front. The advantage of this is that you have no need to make new drawings if you want to translate into a different language. The drawings and the letters written on them will remain the same and one only has to get a translation of the written page, which is normally fairly easy to do.

Farm Surveys

ITDG surveys the "marketplace" to determine what the real problems facing the local community are. This technique has been particularly aimed at African agriculture because most of our experience has been gained in Africa. Now, it is well known that agricultural research in developing countries has very largely been carried out by expatriate scientists over the past 50 years or so. A great deal of this research has been done and most of it has been very carefully conceived by European standards. The work has been very competently executed and it has clarified a great many of the features affecting the agriculture in these countries. Of course, in theory, this research information is of great value to the African small-holder. But if you visit a research station in Africa and then look at the farms around it, you will notice immediately the contrasts; the very high crop yields in the research station, and the relatively very low ones in the farms around. As an example, take the fairly typical case of seed cotton production in Nigeria. A research station in Nigeria can easily obtain 3,000 kilograms per hectare of seed cotton; the farmers' production is nearer 300 kilograms. In the past it has been the custom of the research workers, if they have in fact been questioned about it, to put the blame mostly on the agricultural extension service. The explanation has been that the scientists have produced the information, but the extension worker has failed. He has not put it across to the farmers.

I think we are beginning to see now that the real truth is that in very many cases the sort of research results that are produced are just not capable of being applied to the conditions of life and work on an African small-holding. They are inappropriate for some reason or another—technological, economic, social. Most people by now are surely aware that there has been very limited success in the green revolution, because the new crop varieties need large amounts of fertilizer, they need all sorts of pesticides and other inputs which normally are just not available to the small-holders. And the same sort of thing has applied to much of the agricultural development effort in Africa. So as a prerequisite for the successful application of the results of agricultural research, you must find out a lot about the way of life of the small farmer. And you really do not gain this knowledge by doing a very large, macro-level, social or economic survey. A lot of these surveys have been carried out in recent years; they are valuable in themselves, of course, but they don't throw any light on the basic problem. There is a need for surveys right on the farms—the micro-scale survey. If you are to introduce the correct innovations or improved management techniques, they have to be introduced in the appropriate order; we have seen that it is of no use giving people a crop variety that needs a lot of fertilizer if one has not got a system for distributing the fertilizer. Therefore, innovations must come in the appropriate order and they must be consistent with the actual needs of the rural community.

To find out these needs, really requires a very comprehensive survey. The ITDG publication on survey methods covers the sort of information one needs: just how much labor is available on the farm, how many people are there living in the household, where is this labor used, how much of it goes into crop production, how much into livestock enterprises, how much in the household. Very many African farmers spend part of their time trading or doing some other sort of work and you therefore need to know what is the labor pattern in the target group. You need to know whether the homestead is permanent, whether there is some sort of a shifting cultivation, or whether farmers will move after a few years to a patch of soil which has regained its fertility. You must know which individual within the family makes the decisions relative to agriculture (this may not be done by the farmer himself).

An example has been found in several countries where maize has been introduced, or where the extension service has tried to promote its use. Maize is a very hard crop to grind by the traditional method of pounding, and what happens is that the women complain that it is too hard work and they don't want to do it. So the farmer stops growing it, but it was his wife who made the decision.

You need to know what the land tenure system is. Generally speaking, in Europe a farmer has a fairly stable sort of position on his own farm and he knows he is going to live there for maybe the rest of his life, and perhaps his children and grandchildren after him. So, if he makes some improvements it is well worthwhile.

But if the man has little security, if he may be thrown out next year, or if the land ownership changes from one man to another every few years, then nobody will put in any fixed investments on any large scale. You need to know whether there are any social or communal activities or obligations. It may be the custom in the area that all the farmers work together, perhaps for the harvesting of a particular crop. There may be a religious festival which means that it is no good introducing a new crop which must be harvested at this time because people will not want to be working then. It is very difficult to change this sort of built-in social obligation. So you need to know quite a lot about the farmer.

Additionally, of course, you need the standard agricultural information such as how big is the farm, how big are the fields, are the fields all in one place or does the farmer have to travel several miles between fields, the type of soil, the fertility of the soil, and the trend of the fertility of the soil. In many parts of the north of Nigeria you can meet farmers who say, "My grandfather used to fill six grain stores from this land and now I can only fill two from the same land."—which means that the fertility is declining. You need to know if there is additional land available. It is not much good introducing a new cultivation technique which will allow a farmer to cultivate large areas of land, if there is very little good land available for him to cultivate anyway. You need to know whether there are any soil conservation works which are necessary. You need to know, of course, what is the rainfall pattern and how reliable is the rainfall. This is the most important determining factor in agriculture. You need to know the cropping pattern and to which crops the farmer gives priority. There have been many years of research, again in the north of Nigeria, to help farmers grow more of their main cash crop, which is cotton, and to encourage them to plant it early. If a farmer plants cotton in the month of June, he will get about twice the yield that he does if it is planted, as the farmers do, towards the end of July. Here again it was only realized very recently that the reason farmers don't plant until the end of July is that they are busy working on other crops; they are concerned with their staple food crops. The farmer is going to make sure that he gets enough to eat, even if he doesn't get much spare cash, so he will always plant and weed the land for his food crops before he will even consider planting cotton. This cash crop is just a sort of bonus to him; his is only a partially monetized economy.

You need to know, of course, about the marketing of crops, what is the marketing system, how far do the crops have to travel, are they perishable crops or not. These questions will reveal only part of the information that one really has to find out thoroughly before even starting to do research on innovations like appropriate equipment or techniques.

Certainly there has in the past been quite a lot of research which has been totally inappropriate to the situation of the African farmer. But if you can do the type of survey I have just described, you should find out where the labor peaks occur, that is, at what time of the agricultural year is the demand for labor at its peak. This is probably the point at which the introduction of machinery will be most effective. In the past it has very often been generally assumed that it was some particular task that was imposing a limit to productivity, and it has frequently been said that the solution to the farmers' problems is ox plowing—what they need is ox plows to help with the hard work of preparing the land for cropping. But it has been found that even after the ox plows have been introduced the farmers still do not grow any more, or any better, crops, because ox plowing has not helped them much. It may

have made life easier for the farmer, which is good in its way, but plowing is not really the labor bottleneck. The labor bottleneck comes at the time of weeding the crops, and it is no good planting twice as many crops as he can keep free of weeds.

Demonstration Projects

I shall quickly mention the field projects which we have run. The area in Zambia in which we were involved was one for which there was not very much agricultural information. The project started out with a survey on the lines I have described. They found out that the main problem was transport, but weed control and the harvesting of groundnuts—one of the major crops—were also important. Therefore, once these bottlenecks had been identified, the engineers on the project developed some bullock carts for transport—not used in the area before—and various types of cultivators for killing weeds together with an implement for harvesting groundnuts. In this area of Zambia there were not very many blacksmiths, or rural craftsmen, capable of building such designs. So it was necessary to train the rural craftsmen, but they didn't have even the basic blacksmithing equipment. So, before the project staff could begin training, they had to design a very simple blacksmithing forge and then show the people first how to build it. This just goes to underline how important it is to do the right things in the right order—until the blacksmith has got a forge he can't make any implements, however they are designed.

In the Nigerian project a good deal of the survey data had already been collected by the economics department of the university, and weed control had been identified as the labor bottleneck. The project staff developed implements for applying herbicides, both as a spray and as

39

granules. As a result of this, an interesting technique was developed largely in the university, but with some assistance from the ITDG project. The university agronomists decided it might be an improvement to try to mix the herbicide with the fertilizer granules and simply spread the resulting mixture—in other words, to apply the fertilizer and the herbicide in one operation. A number of machines for the mechanical control of weeds by hoeing were developed by the project together with equipment for harvesting groundnuts and for processing kenaf which is a fiber-producing plant. It is a very tall plant with a thin stem, the fiber being on the outside whilst the inner part is soft. The engineers designed a machine to get rid of this soft part and to separate the fibers which can then be used for making rope.

All the implements, in both projects, have been designed for local construction using the sort of materials which are easily available and employing simple equipment. It is important to stress the need to use easily available materials; as soon as one gets away from the cities in, say, Nigeria, there is no hope of building an implement which is made out of what we in Europe would consider a standard sort of engineering material such as flat strips of steel or iron. These things are just not available outside the large cities. The only sort of steel which is available is that which is used by builders—steel water pipes and steel rods for reinforcing concrete. So you have to design, or very often redesign, machines to be made out of water pipe and reinforcing rod. In the Nigeria project we made the prototypes in the project workshop and then commissioned from a local craftsman half-a-dozen of each machine for testing by farmers.

In this way the local craftsman learned how to make this sort of machine and got paid for making it. If at the end, the farmers

decide they like the machine, there will be a local manufacturer already trained. What has happened in the past in some cases is that a man has developed a machine perhaps at the local university, it has been shown to the farmers who say, "Yes, a very nice machine, where can we buy it, how much will it cost us?" But if nobody has really looked hard at establishing proper manufacturing capability then the only reply can be, "We don't know where you can buy it, nor do we know how much it will cost," and of course the farmers lose interest. Many schemes have tried partially to overcome this by bringing the black-smiths into a central training establishment, but very often a blacksmith doesn't want to come—who is going to keep him alive, who is going to feed him and his family when he stops work and goes away for training? This was the reason we got local craftsmen to build the machines in their own workshops on the spot.

Consultation

Our involvement with agricultural consultation is only just beginning, although the other speakers in this series mention many other topics in which assistance and advice has been given to enable schemes appropriate to the local situation to be planned and executed. We have, however, undertaken a consultation for a bauxite mining company in Guyana, which operates in a very underdeveloped region. The company is already considering its responsibility to the local workers at the time when the bauxite supply runs out in about five years' time. There will then be something like 5,000 people without jobs in this isolated region. They are, however, already well established there, with their own homes and their children growing up in a living community. To move 5,000 people when the bauxite is exhausted is not only expensive but also, in the light of other rehabilitation schemes, stands a very high chance of being socially disastrous.

The company therefore is examining what possibilities there are for this community, and engaged us as consultants to look at, among other things, the feasibility of starting some sort of agro-based industries. They are thinking of growing cassava which is a crop with a very high starch content. The mining industry uses a great deal of starch in processing bauxite so the agriculture and the industry processing cassava into starch can be combined and then supply the mining industry. Another agro-industry that is being considered is based on goat rearing. The goats can produce meat and milk and could give rise to additional processing industries manufacturing fertilizer, stock feed, glue, and other such products.

Tools for Agriculture

I hope that by now I have given a fairly comprehensive account of the way that a European-based group can assist in the improvement of the agriculture of developing countries. Although the problem is vast we have identified for ourselves a few highly specialized tasks. Perhaps the most important aspect of the work is in the mobilization of knowledge that is simply not available at the grass roots where the innovations must take place. So the information that we handle is of necessity both very generalized, in the sense that it is collected from widely divergent regions and types of farming, and highly specialized, in the sense that all the innovations must be carefully selected to be as appropriate as possible to the agricultural parameters and social customs of the area where they are to be introduced. The process that connects these two aspects and enables the specialized selection to be made is, of course, the farm-level survey.

But in order to give a clearer understanding of the type of equipment we are talking about, I want to give some examples. Now there are two ways that you might react to them. You might think that technically the designs are extremely simple and that it is hardly worth a highly qualified engineer's time to be bothered with such seemingly minor improvements to crude traditional implements. Or you might understand, particularly if you have had field experience in a developing country, that the difficult part is not the design work but it is identifying precisely what type of equipment will be of use to the farmers, ensuring that it can be manufactured locally and, above all, seeing that it is actually used.

I have grouped the illustrations that follow roughly in the sequence of operations that must be performed on the land and the crop. Clearly this is no attempt to describe all the operations that are necessary or that might be desirable. Nor does it mention more than a tiny sample of equipment that might be appropriate in different circumstances. Moreover, the examples are not necessarily developments by, or any work to do with, the Intermediate Technology Development Group. They are simply a random selection of designs that have originated from a number of countries.

Power

The availability of suitably cheap and reliable forms of power to apply to agricultural operations is obviously important and is the subject of much research. I have already mentioned one approach—the "snail" stationary power unit. A slightly more traditional unit was employed by the university in Uganda—a small, very simple, and low-cost tractor for African farmers. It uses quite a lot of imported parts, principally the engine, gearbox, axles, and steering gear. But the main

41

Photo 3.1. Indian bullock yoke.

frame was designed and built locally and all the assembly was local. A basic limitation of small tractors seems to be their light weight and small wheels together with a relatively high cost. This design did not develop much power and has not really been a great success.

Many more farmers in the world, of course, use animals than use tractors, although a calculation of the actual energy consumed is not so easy. Quite considerable improvements in the animals' output can be made through attention to breeding,

animal management, and the design of the equipment they are to use. Photo 3.1 gives an idea of the sort of yoke commonly fitted in India to the necks of these bullocks. It is not very comfortable and therefore not very efficient. In many areas there are improved local designs that, in the case of the cattle shown here, would fit round the hump and enable more of the body weight to be utilized for tractive effort.

Furthermore, we have demonstrated to farmers in Nigeria how to work with just one bull. As a matter of interest the bulls in

42

Photo 3.2. Simple moldboard plow.

that country are very tame and the use of bullocks or oxen in Nigeria is quite rare. At present most of the farmers in Nigeria work their cattle in pairs even for light operations. This, of course, is very wasteful, for if the man can use one bull at a time he can either let one of them rest while the other is working or he can buy another implement and use the two implements together. This would be an example of a very simple, though important, improvement.

Cultivation

One of the chief tasks for which power is required in agriculture is cultivation, or the preparation of the ground for planting. Photo 3.2 is an example of a very simple sort of moldboard plow that is being

Photo 3.3. Improved plow for crops grown on ridges.

demonstrated in India. Although it seems a very basic implement, it is itself a considerable improvement on the type of plow that is very widely used in India. It has been properly designed—you can see a range of adjustment holes to vary the angle of the plowshare—and well made. The material that the main part of the plow, shown in photo 3.3, is made from is water pipe, which is a very cheap and useful material. The plow is an improved type for use with crops grown on ridges and is again made in India.

There are other types of cultivation equipment than the plow. Most people associate disc harrows with sophisticated Western farming and large tractors but they can be made even for use with bullocks (photo 3.4). The performance of this machine is not as effective as those used with tractors, but it does a good job in the circumstances. Photo 3.5 is just a very simple cultivator being exhibited at an agricultural show. It is made very largely from wood with only a small amount of steel used principally for the tines, but again it can do a good job.

Photo 3.4. Disc harrow with bullocks.

Photo 3.5. Simple cultivator.

Photo 3.6. Scoop for earth moving.

Photo 3.7. Simple planter.

Levelling

Soil erosion can be a serious hazard to agriculture and to the morale and livelihood of the people trying to support themselves on it. I have seen many cases in East Africa of rains washing away three or four centimeters of soil complete with the newly planted seeds. The only effective remedy involves lots of hard work in digging a drainage system or in reshaping the land's contours. In many areas it is necessary to move a lot of soil in order to form fairly level terraces. The European way to do this is to use a tractor or a purpose-built grader. But these machines have to be imported, are very expensive, and need very skillful operation. All too frequently they are liable to break down miles from a skilled mechanic and well-equipped workshop. One way to overcome this problem and simplify the operation is shown in photo 3.6. It is just a simple implement for scooping the soil that is pulled

45

Photo 3.8. Partly mechanized planter.

behind a pair of bullocks. Again, it doesn't perform as well as the latest sophisticated machine, but it is very easy to make, requires no foreign exchange to purchase it, anybody can repair it, and an untrained man can use it.

Planting

After the land has been cultivated and levelled, one has to plant the crop and photo 3.7 shows the very simplest sort of mechanized planting arrangement. You can see that it consists only of a sort of tine which is pulled through the soil by a bullock. The seed is dropped down a tube into the furrow opened up by the tine. This simple machine represents a considerable advance over hand-seeding; it is much faster and much easier. This type of implement can be developed further by having a similar device sowing four rows at a time, although it takes quite a lot of skill to drop the seeds down so that just a few flow into each tube. With a bit more development you can partially mechanize the system as is shown in photo 3.8. A hopper for the

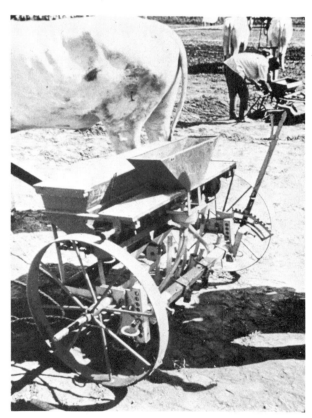

Photo 3.9. Seed-and-fertilizer drill.

46

Photo 3.10. Animal-drawn hoe.

seeds and the tubes that direct them into the ground have been attached to what is basically an indigenous traditional plow. The additions are made out of steel; the plow is wood.

The same research station in India that developed the previous example is now moving towards a combined seed-and-fertilizer drill (photo 3.9). You can see that it is not quite as expensive to produce this

Photo 3.11. Weeding attachment fitted to ridging plow.

Photo 3.12. Hand-operated hoe.

design as the Western type of seed drill, which meters the seeds out separately for each row. In this case there is only one expensive part to the machine—the part at the top which controls the flow of the seed into a funnel and down through four tubes into four rows in the ground.

Weed Control

The next operation is that of weed control and photo 3.10 shows a conventional type of implement to be pulled by cattle for hoeing out weeds. This particular example was made on our Nigerian project and is nearly all constructed of wood which cuts the cost considerably. It can be made with no more tools than the average village blacksmith already possesses.

In the north of Nigeria most farmers already have a ridging plow but photo 3.11 shows a weeding attachment fitted to one. The plow consists of the handles, the plowshare, and the drawbar which are

Photo 3.13. Ridge crop weeder.

darker in the photograph. The lighter shaded part is the weed control attachment which simply does some hoeing—removing weeds at the same time that the plow is pulled along. Although very cheap, it is quite effective in the right conditions.

A very simple specialized hoe is shown in photo 3.12. This requires no animal power—it is merely pushed along by hand. But the wheels make it much faster and less tiring to use than an ordinary hoe that is just held in the hand. And it is a lot cheaper to buy and use than one powered even by oxen, and certainly very much cheaper than anything powered by a tractor. Photo 3.13 is another attachment, but it is specialized in the sense that it is used for weeding crops grown on ridges. The attachment is designed to be used with the standard ox-drawn toolframe shown. A very similar design of machine made on the same principle, but for use with a tractor, costs about $1,200 and will weed three rows. For this one we managed to reduce the price to about $20 to enable a man and a pair of oxen to weed one row. Further development produced a weeder that could tackle two rows at a time and a slightly different design capable of weeding crops standing about one meter high.

Pesticides and Fertilizers

In the current state of the agricultural art it can be shown without doubt that certain practices, which are derived from the European approach to agriculture, lead to increases in output. A classical example is the benefits gained from sowing seeds in rows rather than broadcasting them indiscriminately. Similar benefits over traditional agriculture are attainable from the use of chemical inputs, particularly pesticides and fertilizers, and the small farmer has generally recognized this and is eager to use them. Without going into what at the moment are the more theoretical arguments about the wisdom of basing

whole agricultural systems on chemical, as opposed perhaps to biological, inputs or into the arguments about the continuing availability of such inputs, one cannot advise that the small farmer, should not benefit from them. His margins of safety, profit, and livelihood are too small. But with a clear awareness of the dangers of energy-intensive farming and with a determination to see that the small farmer has access to equipment he can afford, one can, and I believe should, make available to him appropriate equipment for the application of pesticides and fertilizers.

There has been a great deal of thought given to machinery for pesticide application. The knapsack type of sprayer, carried on a man's back, is very familiar. Two men with hand-pumped equipment on their backs can spray about half the width that a tractor unit can. But the equipment cost is approximately $50 whereas the tractor and its sprayer would cost about 100 times that amount. Engine-driven knapsack dusters are even more impressive; and they can be designed as a two-man unit. One man carries the reservoir, pump, and a reel dusting tube, the far end of which is taken by the other man. Together they can cover a width of about nine meters which is as wide as tractor-drawn dusters will cover. And the equipment is in many cases much better suited to use in small fields.

Threshing and Processing

After all the operations have been carried out on the land and the growing crop, and it has been harvested, it needs further processing before it can be used or sold. Many crops require threshing. In many countries the crop is laid out in a big circle and bullocks pull a sledge round and round in a circle while walking over the crop. Eventually most of the grain will fall to the bottom of the circular heap; about 60 percent of it will be recovered although the remainder will be wasted. Photo 3.14

Photo 3.14. Mechanical thresher with an output of 100 kilograms per hour.

Photo 3.15. Large thresher used for contract work.

Photo 3.16. Device for stripping kenaf fibers.

shows an Indian improvement. This higher level of technology is an engine-powered thresher with an output of about 100 kilograms per hour. It is a fairly cheap machine—certainly much cheaper and more appropriate to their conditions than a combine harvester. Photo 3.15 is another thresher made in India. This one is mounted on large wheels so that it can be towed by a tractor over all sorts of ground and it has an even larger output. It is very suitable for contract work. Photo 3.16 is some simple equipment for the processing of kenaf which I mentioned earlier—its purpose is to strip the fibers from the plant. You can see that many bicycle components are used in its construction, because they are cheap and easily available in Nigeria where it was made. It even utilizes a pair of discarded valve springs from a tractor engine which are also easily available in most places in Africa.

Manufacture

I referred earlier to the importance of the design that we made of a very cheap metal-bending machine. This is a vital piece of equipment for the local craftsman to have if he is to be able to fabricate the machinery designs we have just been considering. At a slightly more advanced level of technology is a fly press that has been developed and built by the owner of a small metalworking factory in India. A fly press is normally a fairly small piece of equipment, but this is a very large one requiring six men for its operation. They have to run round in a circle, screwing the stamping part down, in order to bend hot pieces of steel. The factory owner claims that it is equivalent to a 100-ton power press and that for the occasional use he has for it, it performs more than satisfactorily.

Conclusion

In this chapter I have made no mention of many of the wider issues such as land reform or financing. This is not because they are not vital, but because there are many organizations and programs already committed to these aspects. One of

51

the most difficult tasks that the agricultural project has faced was to define a way in which a very small amount of money and manpower based in Britain could do something practical and useful. The most important gap that we could see, and that we felt we could fill, lay in the provision of knowledge about different levels of technology. In order to fill this gap we have had of necessity to become involved in economic and social surveys. We feel that the reason that this was necessary was because nearly all the surveys that had been done before had no technological component. That is to say that the people carrying them out had insufficient experience of technology and the possibilities and limitations of agricultural equipment, to look at the problem from the farmer's point of view. One might almost say that one can't ask the right questions until one knows at least some of the answers. But we don't think it strange that we have had to become involved in nonengineering disciplines, since we don't think that engineers acting alone can make much of a worthwhile contribution either.

Chapter Four
Intermediate Technology Building
by John P. Parry

Man was once a cave dweller. In the Ice Age this provided a form of accommodation which ensured the survival of the species. Even today it is hard to improve upon the four main advantages of a good dry cave:

Moderate ambient temperatures	Caves have excellent thermal insulation and can be made draft free. The temperature only changes over a narrow range between day and night, winter and summer.
Low maintenance costs	A good cave incurs almost no structural damage in use, is therefore virtually maintenance free and lasts for thousands of years.
Hygienic conditions	If looked after and swept out regularly, a cave can be kept free from most disease-carrying vermin. It is important to throw bones well clear of the entrance and also to bury the dead as far away as possible.
Security	Single-entrance caves are ideal for securing against thieves and wild animals.

Caves are, however, in very short supply and are badly situated in mountainous areas away from flat agricultural land. The nearest existing equivalent to a cave built artificially on flat ground is the Scottish "black-house," a hollow mound of earth and rock with a thick roof-covering of natural vegetation.

As the earth's temperature rose and the ice receded, man was able to move out of caves in the hills and change from a hunting/gathering economy to farming.

Shelter was still needed, especially during the winter and at night, and people became enormously resourceful in building dwellings out of naturally occurring materials. By assembling materials which could be found around them into simple buildings, people were able to survive in areas of semidesert, on grasslands and arctic tundra, in forests, river floodplains, and even in river estuaries with houses actually built on stilts in the water.

One of the most rudimentary of dwellings still being built today is the "sulap," a simple lean-to made out of bark from trees and sticks, accommodating one or two persons, which the nomadic Punan people of Borneo use for overnight shelters. (I have spent nights in sulaps on many occasions and can describe the experience as most uncomfortable. The simple structure keeps out direct rainfall but provides little protection against leeches, mosquitoes, or the damp night air of the rain forest.)

Building Materials

In different parts of the world all kinds of materials have been used for building, most of them impermanent.

People living in rural areas initially built with natural materials such as mud, bamboo, logs and sticks, tree bark, rocks, leaves, grass, reeds, coral, seashells, and snow. These materials can be gathered and put to use with little further processing.

Taking development a stage further, other building materials are made by *converting* naturally occurring resources; planks cut out of wood, building blocks chipped from soft stone, and slate tiles cut from hard layers of sedimentary rock. Even animal skins are made up into tents by nomadic herdsmen and so constitute a building material.

The ultimate stage in the evolution of basic building materials is the deliberate manufacture of permanent building components from soils and aggregates. These new components are bricks, blocks, tiles, pipes, and cements which require for their manufacture a prior input of capital and the exercise of special skills.

In recent years, the waste products of civilization have also been used as building materials by shantytown dwellers: kerosene tins, hammered flat, are used for roofing tiles; old packing cases, and even pieces of old tires, make rough walls. In Australia the walls of huts have been built out of empty beer bottles which occur in large quantities in outlying townships.

The facility to find and adapt materials to construct suitable shelters could well be an inborn characteristic of our species—a mechanism for survival just as it is with many species of mammal, bird, reptile, fish, and crustacean. Perhaps we can see this instinct in the behavior of children who all seem to take pleasure in building "dens" or shelters.

What is apparent is that people of all races have always been incredibly resourceful in their use of whatever materials they can find to construct their buildings. The results have been remarkable for they have enabled life to go on, children to be born and raised, and men and women to rest, in conditions so extreme that unprotected livestock and even wild animals have died.

But there are, however, grave drawbacks to building with materials which are perishable and irregular in shape, some of which are obvious and others not so obvious.

Most dwellings built of natural materials have a limited expectation of life, sometimes three years or less. Moreover, there is a heavy burden of *maintenance work* to keep the structure habitable. In an economy in periods of the year when all a family's waking hours must be spent in production of food, it is an unwelcome additional burden to have to rebuild part of the house after a heavy rainstorm. Houses which *depreciate* in value in time produce less stable economic attitudes than do houses built of permanent materials which tend to increase in value in time. If a man has a valuable piece of property he will tend to persevere with his means of livelihood even after a spell of bad luck—a poor crop, a gambling loss, etc. But if the house is worth virtually nothing it is easier to leave it behind and seek better luck in the city. Their lack of firm rural roots tends to swell the numbers of people drifting to the cities in developing countries, causing the dreadful problem of urban overcrowding and squalor.

There is a danger of infestation by disease-carrying insects in most tropical structures which are built of bark, leaves, or unpainted wood. Fleas and bedbugs spread plague, etc. Irregular structures are difficult to screen against mosquitoes, carriers of malaria and elephantiasis. It commonly occurs that people are forced to abandon an infested dwelling even though its structure is still sound.

Indigenous dwellings are designed as far as possible to overcome the worst environmental problems of heat, cold, rain, wind, dust, and harmful animals and insects. Their shortcomings in these respects are largely caused by the unsatisfactory properties of the *building materials*, not by the design of the dwelling.

Environment

There is in fact no universal solution to a building design. In some places it is desirable to keep the wind out, elsewhere it is desirable to let it in; similarly with sunlight. In some places people may want roofs to be flat providing a high place to sleep during hot nights or an area to dry foodstuffs; elsewhere the roof must be steeply sloped and strong to prevent damage from snowfall. Some cultures have small family units of four or five people of two generations; others expect all living family generations and close relatives to live together in households of 20 or 30. In Borneo where some people dwell in the sulap as described earlier, other tribes go to the opposite extreme and build huge longhouses where several hundred people, the whole population of a village, all live under one roof. To try to standardize the building systems even of a single country could tamper with cultures. Community life could be altered with unknown consequences. Even in a single country, climates can vary within a few miles, posing many different problems. In these areas the

seasonal patterns are also different and the various types of agriculture have different associated needs for storage of food and tools.

To a worrying degree the application of modern building technology to the developing world has resulted in a deterioration of standards of amenity. Many low-cost housing schemes have resulted in the establishment of dwellings which are culturally and climatically inappropriate. In contrast with the variety of designs of indigenous dwellings which take into account night and day temperature differentials, prevailing winds, need for storage of goods, family structure, and number of wives, the European-inspired house is usually a standard concrete box with an asbestos-cement or corrugated iron roof. It is not surprising that there have been instances where villages have been moved as a result of road or irrigation schemes and rather than live in the new houses provided, they have built their traditional structures alongside them. It is painful to contemplate what a tragic waste of scarce resources such mistakes represent in a poor country.

The apparent material superiority of European and American industrialization has made their methods the obvious model for the undeveloped countries. By a combination of imposition and unquestioning acceptance, Western building technology has already dominated the skylines of the cities of the Third World—and even in rural areas the architect from the underdeveloped country, having been trained in Britain, Germany, or the United States, builds to fit his own countrymen into houses conceived for another culture, economy, and climate altogether.

The compromise is fairly obvious. Providing that the facilities to manufacture permanent basic building materials are available, most of the "design" can be left to the individuals who have already demonstrated incredible achievements in their use of unsatisfactory building materials in the past. Surely they can be given the discretion to use far better materials once they are made available.

Skills

But how about the artisan skills of joinery, bricklaying, tilehanging, and plastering? The overseas work of ITDG is concerned mainly with rural areas and the most important approach here is that of encouraging self-sufficiency. While it takes years of apprenticeship to learn artisan skills to achieve the standard of building expected in European housing, how can we expect to train the whole rural population to the same standards? We cannot nor should we even try to when the other skills of animal husbandry, maintenance of implements, planting, tending and harvesting crops are vital for survival. Therefore, we should do no more than teach people a few basic tricks-of-the-trade which will enable them to adapt their existing building skills to make use of the better materials. These techniques should be demonstrated at the plants where the materials are made; the building-to-survive instinct should do the rest.

Where government building projects are involved for housing, schools, and clinics in rural areas, it is of course necessary to adopt a more systematic approach to the design and construction of the buildings along appropriate lines. In this case, instead of reaching for the textbook, the professional architect should reevaluate the indigenous system of building in that local area and develop ways of upgrading it by substituting unsatisfactory materials with permanent ones. Only when it is clear that the nature of the available traditional material is the major influence on the design of the indigenous structure, should the building methods be radically

altered to accept permanent and better materials.

Technical Assistance

The role of the Intermediate Technology Development Group is therefore primarily to provide three forms of assistance to developing countries' building activity:

1. To encourage the setting up of appropriate manufacturing facilities for permanent building materials—bricks, pipes, cements, tiles and roofing sheets—and with simple windows and doors as a lower priority
2. To set up appropriate procedures for training rural people in manufacturing methods to produce the building materials and for demonstrating how they can be put into use in upgrading traditional building methods
3. Where government departments are building in rural areas, to encourage the designers and contractors to take advantage of the climatically appropriate local designs, while incorporating, where possible, improved locally produced building materials.

The key word, therefore, is "appropriate" and this applies equally to the method of manufacturing materials. A good specialist/technologist would have little difficulty in designing a conventional brick and tile factory to use the clays of rural Africa and Asia. From his catalogs of machinery would come a range of proprietary excavators, conveyors, grinding mills, brick presses, dryers, and kilns with a capital investment of at least $2 million, amounting to $40,000 per workplace. Such a prestigious project would be all very well as a showpiece for visiting ambassadors but even a country as small as Ghana would need a hundred such factories to

Photo 4.1. Asokwa brickworks (Ghana).

provide the bricks for its housing needs and where would the capital be found for such a huge investment program?

Small Brickworks

I mention Ghana because this is where in 1973 two new brick factories started operation having been built along the lines advocated by ITDG. The project to design and develop an appropriate solution to small-scale brickmaking was financed under British Technical Aid and the building of the works carried out by the Ghanaian Building and Road Research Institute.

The first works was built at Asokwa, a small village 60 kilometers south of Kumasi. It produces about 10,000 bricks a week and employs 26 men from the village. It cost about $20,000 in total, which represents an investment of under $800 per workplace—one-fiftieth of the capital per workplace of a conventional

factory, the output of which would only have been 30 times larger. But there are several other differences which are even more important.

Photo 4.2. Preparing clots and molding the bricks.

Photo 4.3. Tipping bricks out of molds onto drying pallets.

Photo 4.4. Stacking wet bricks on drying racks.

Photo 4.5. Locally made clay mixer.

Photo 4.6. Kiln opened after firing completed.

1. The capital cost to build the intermediate-technology factory involved under 10 percent imports. The rest was obtained locally. By comparison, 70 percent of the capital cost of the conventional factory would have involved imported materials and equipment. The significance of this for a country starved by foreign exchange is profound.

2. The Asokwa factory uses natural air to dry the bricks after they have been shaped and the kiln burns local firewood. The only imported source of energy consumed is for a small 10 h.p. diesel engine driving the clay mixer. This uses 10 tons of oil for a production of a million bricks. In a conventional modern factory, however, 150 tons of oil would be required for the kiln alone, 15 times as much, and a substantial input of electric power is also needed. The Asokwa factory on the other hand operates without an electric power supply.

3. At Asokwa all the necessary skills of brickmaking were gained by ordinary villagers after only eight weeks' training. The technology concentrated on fairly simple skills of handmolding, stacking the bricks in the kiln, and operating the simple motor-driven mixer. In the event of mechanical breakdown, the tools and equipment were types familiar to local car mechanics who also service the small maize-grinding mills which operate in the same village. By contrast, the modern factory would require at least four highly skilled technicians, expatriate or local people trained in Europe, to operate the complicated

kiln, dryer, and the mixing and pressing machinery.

Based on the success of the first factory at Asokwa, the government of Ghana instructed a second one to be built at Ankaful near Cape Coast. This one was successfully brought into production and a third and fourth works are being built elsewhere, now without any need to call for help from myself or my colleagues. The technology is now disseminating without outside assistance, which is another important aspect of an intermediate technology.

By employing an intermediate technology it was possible to bring production into rural areas of Ghana and provide work for unemployed people in these locations. A modern factory would have had to be built in the city to have any chance of obtaining the more sophisticated labor force required—and would have further added to the drain on the country's resources and to the extreme problem of urban overcrowding.

Cement Substitutes

More recently, the advocates of appropriate technology have been giving great attention to the possibilities of cement substitutes for developing countries.

Cement technology is high technology—if for no other reason, the need to attain a process temperature of 1,500°C. requires the application of heat-resisting linings and sophisticated control procedures. Once the decision to apply such a technology is made, it is the obvious course to operate on a large scale to spread the cost of the expensive instrumentation and hardware. Rather than attempt to scale down and simplify the manufacture of cement, the group considered that it would be better to explore ways of developing the

alternative technology of lime-pozzolana mortars as a substitute for portland cement. Process temperatures for lime production are little over half that for cement and can therefore be adapted to simpler technologies. There exists already an indigenous pozzolana technology which is widespread. The combination of slaked lime and "surkhi" (waste burnt clay products ground down to a powder) is used both as a mortar and to produce a concrete. The Sri-Ram Institute in Delhi is developing an improved technology for the production of pozzolanas using a fluidized bed. The best prospect for development of an appropriate *rural* technology, however, is probably in the improvement of the vertical shaft kiln which exists in one form or another in Sri Lanka, Colombia, Ethiopia, Somalia, and no doubt, many other developing countries as well.

These kilns need to be made more efficient in their use of fuel, and simple aids need to be devised to ease the backbreaking labor of loading the limestone and extracting and slaking the lime. A further improvement is to help the limeburners to obtain better control of quality, by making available a robust and simple means to measure temperature. Control of the Asokwa brick kiln is done by measuring the shrinkage of the stack of bricks. The extent of "heat work" done by the fire to complete the ceramic process can be accurately monitored by the shrinkage of most clays. For lime burning it is more necessary to have some measure of the actual temperature. By encouraging the practice of combining the lime with a suitable pozzolana—a technology which is much less widespread—ITDG will be able to assist rural areas to be more independent of expensive proprietary cements which represents a further step towards self-sufficiency.

Domed Roofs

A further technology which the Building and Building Materials Panel is seeking to revive and improve, is that of the dome and the barrel vault as an alternative form of roofing. Great strides are being made in dome building by Hassan Fathy in Egypt who has the benefit of a traditional village building method to start from. Domes and vaults are alien to the people of most other cultures, however, and it is therefore our thinking that for these to be used spontaneously, a special brick shape needs to be evolved which can quickly and with little skill be built up into a curved roof. The most promising shape is the simple barrel vault which can start from ground level or be placed on a rectangular room. Our thinking so far is that the basic brick shape should be a modified hexagon which would fit together and also produce a self-locking tie with a second skin of bricks laid on top. There would be a thin layer of mortar in between. The material would therefore be halfway between a brick and a tile but would have the advantage of being a good thermal insulator. More important than this is the prospect of overcoming the costly and time-consuming work of building a permanent timber frame which is the biggest obstruction to the use of roofing tiles in developing countries.

When the people in the developing world can build themselves good brick and tile houses with the masonry bonded by pozzolanic mortars, all of which materials they have made themselves, we will have taken a major step towards reattaining the standards of maintenance, insulation, and permanence of that excellent but scarce dwelling, the Ice Age cave.

Chapter Five

Energy in Rural Areas:
An Intermediate Technology Approach

by Peter D. Dunn

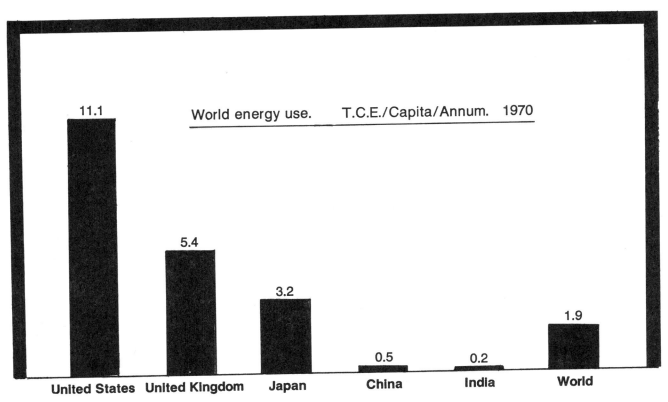

Figure 5.1.

Energy is a basic need with our modern technological civilization and perhaps a few statistics will be worthwhile. We currently use about 6.8×10^9 tons of coal equivalent a year in the world. I think this is the 1970 figure, but it is much the same still. One t.c.e. represents one ton of coal or its energy equivalent and is a useful sort of unit because most of us can visualize a ton of coal. If we divide this figure by the number of people in the world, the average comes out at about 1.9 t.c.e. per head per year. Thus we are all us-ing something like two tons of coal a year each on average. A rather useful conver-sion factor is that a ton of coal is about a kilowatt, so that on average each of us re-quires the equivalent of a two-bar electric fire to support us. This, of course, is not a true picture; energy is not distributed really very equitably. Figure 5.1 shows how it is spread out. In the United States they use something like 11 or 12 tons of coal a year; we in Europe use something like half that and in the developing world it is much more like half-a-ton. In fact, we can say that of the 3.8 billion people in the world, one billion of us use about five tons each a year giving 5×10^9 t.c.e. energy consumption while the other 2.8 billion people use only one-half each year giving 1.4×10^9 t.c.e.

Now you may well say, so what; what has that got to do with development? But there is a rather interesting correlation between standard of living and energy use. This merely means that in a technological civilization we use a lot of equipment and tools which require energy to build them. The graph of energy per head per year plotted as a function of gross national product is roughly a straight line (figure 5.2).

The United States is at the top, most of us in Europe are around about the middle, but the developing countries are clustered down at the bottom end. So there is clearly some relation between physical well-being and energy, and therefore it is important that energy should receive attention.

Supposing we all raise ourselves to the standard of the United States, as everyone is trying to. Immediately our per capita energy consumption would go up by a factor of six since the United States uses about 12 t.c.e. per person per year and we now average less than 2. Secondly, by the year 2000 the world's population will perhaps be doubled. That would give us an overall factor of 12. The drain on resources would be enormous; other problems would follow including the degradation of the environ-

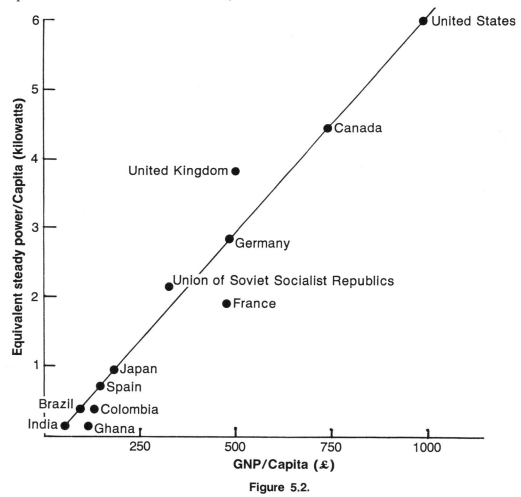

Figure 5.2.

ment and problems of pollution. So I think our long-term energy solution should be to develop a way of life which is energy conserving. Looking at the situation very simply, we really do waste a lot of energy. For example, a quarter of America's energy goes into transport, principally cars taking people to places they don't really want to go at something like 20 percent efficiency; another quarter is burned in power stations at 30 percent efficiency; the rest heats up the atmosphere and water—a great part of which could be conserved by total energy and other schemes. Although this argument is very generalized, I am merely saying that we should look more critically at our way of life and see how we can conserve energy. Even so, we have to develop new forms of energy, that are preferably renewable, within the context of a better environment and a consideration of pollution and safety. These conclusions, although very obvious, do not greatly help the developing countries. Nevertheless, I think one should make these remarks because it is misleading to consider that the whole population can be raised to current United States energy standards; it just is not possible in an environmental or ecological sense. And it should not be necesssary either; we should be considering the type of approach depicted in figure 5.3.

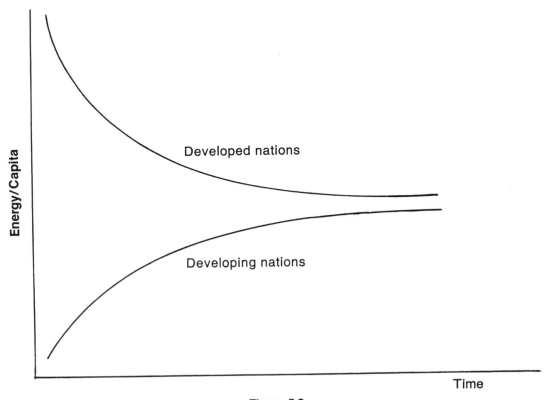

Figure 5.3.

Energy Needs in Rural Areas

Transport, e.g., small vehicles and boats

Agricultural Machinery, e.g., two-wheeled tractors

Crop Processing, e.g., milling

Water Pumping

Small Industries, e.g., workshop equipment

Electricity Generation, e.g., hospitals and schools

Domestic, e.g., cooking

Figure 5.4.

Energy Requirements

Why do people in developing countries need energy? What do they do with it? In developing countries most people live in the rural areas. Thus it is the energy requirement in rural areas at which we should be looking. The sort of things that they need energy for are transport, such as small vehicles or boats; agricultural machinery, as an example, the two-wheel tractor; processing of crops, milling, grinding, winnowing, and similar activities; pumping water for irrigation; setting up of small industries, powering equipment, lathes, circular saws, grindstones; generation of electricity for things like hospitals and schools; and the provision of heat for domestic use, particularly cooking. These energy requirements are summarized in figure 5.4. What we engineers should do in developing countries, then, is to look for a suitable source of energy for these activities and an appropriate means of conversion.

As engineers we ought to consider the criteria against which our solutions are going to be measured. The first thing we are going to ask is: How much power do we need? Then we are going to say: Should it be continuous, does it matter if the supply stops occasionally? Obviously, if you are grinding corn or pumping water it does not; if you are watching your favorite television program, it is quite important.

Cost is a very important second question, and cost, of course, has the usual three components: the initial cost and its amortized value over life, the cost of fuel, and the cost of maintenance. The third thing we are going to ask is: How complex is our solution; can it be made locally? If it is made locally it is particularly suitable for local repair which is a very important aspect. Fourthly: What about maintenance and availability of spares? As those of you who have been in developing countries will know, a tremendous amount of equipment stands idle because it cannot be maintained when spares are not available.

Lastly: How long will it last? And this does not mean in a European laboratory but under the actual conditions when people might put sand in the oil by mistake and make other human errors. In practice there will be a lot of boundary conditions that one does not necessarily meet in Western countries.

Sources of Energy

Having described what we want the energy for, and having outlined our criteria, let us now look at sources of energy and secondly at means of conversion. Figure 5.5 tabulates in a simple way some of the energy sources that are open to us.

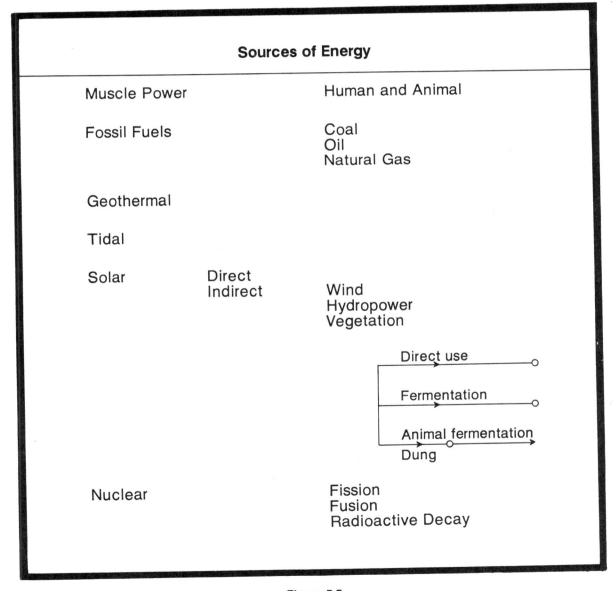

Figure 5.5.

A very convenient source of energy is human muscle. In fact, a good galley slave used to produce about a tenth of a horsepower for an eight-hour day and most of us can do that too. If you work 200 days a year, you can produce about 160 horsepower-hours a year or possibly 120 kilowatt hours, to use a more acceptable unit. Therefore, human and animal muscle power are very important sources of energy.

Fossil fuels, coal, oil, and natural gas are obviously very important. I have listed geothermal energy for completeness. It is not normally very important, unless you happen to live in New Zealand, Iceland, or even northern Italy, where the railway system is run on electricity generated by geothermal power. But there are certain areas in the developing world like Mexico where there is geothermal power which we should use. For tidal energy we clearly need to have a coastline which has to be tidal. But it can be used in certain special circumstances. Solar energy can be divided into direct solar and indirect solar sources. By direct I mean straight sunlight; indirect means the effects of the sun, that is, wind, hydropower, and vegetation. Vegetation you can burn directly as in burning wood, or you can ferment it or feed it to animals, and collect the dung and ferment that. So vegetation can give you various fuels. Nuclear I put at the end—fission, fusion, radioactive decay—just to show that I have not forgotten them. But really they are not applicable in this particular context and, some might argue, in any other.

Conversion Alternatives

Consider methods of energy conversion. It is interesting that the installed capacity in American cars is 20 billion horsepower. In fact, if all Ameri-

cans get in their cars and simultaneously accelerate, the electricity generated by their dynamos would be equal to the electricity generated by all American power stations.

Human power, as noted earlier, is not, as many think, an insignificant source of energy. Assuming that there are approximately four billion (4×10^9) people in the world and that two out of four people are capable of physical work at a rate of one-tenth horsepower, the human "installed capacity" is about 2×10^8 horsepower. While this figure represents only one-hundredth of the installed capacity of American motorcars, it still represents considerable effort. Therefore, from a purely theoretical point of view, it appears desirable that we should find ways to unleash this abundance of "free" energy in both developing and developed countries.

Were we to make a little matrix formed by multiplying the sources of energy by the methods of conversion, we would clearly get quite a lot of possibilities. Some of these we would eliminate because their power range is not suitable for our purpose. It is rather interesting to look at things from the point of view of power range, and this I have done in figure 5.7. What I have plotted horizontally is power range in kilowatts, from 0.01—that is 10 watts—up to about 1,000 megawatts. A man can work, as I have said, at about a tenth of a horsepower—75 watts—for quite a long time. A horse is shown as generating something less than a horsepower and a bullock as generating a little less than that. Petrol engines go up to 100 kilowatts or so, and the same limit is found with high-speed diesel engines. Low-speed diesels of course go up to very big powers. The Humphrey pump goes up to about 10 kilowatts. The gas turbine starts in the megawatt range and goes up to the multi-megawatt range; the reciprocating steam engine ranges from a few kilowatts up to a few

---

Methods of Energy Conversion

Muscle Power   Man
                                Animals

Internal Combustion Engines

      Reciprocating   Gasoline — spark ignition
                                Diesel — compression ignition
                                Humphrey — water piston

      Rotating   Gas turbine

Heat Engines

      Vapor (Rankine) reciprocating* — steam engine
                                rotating — steam turbine
      Gas   (Stirling) reciprocating*
                                (Brayton) rotating — gas turbine
      Electron gas   Thermionic
                                Thermoelectric

Electromagnetic Radiation

      Photo devices

Hydraulic Engines

      Wheels, screws, buckets
      Turbines

Wind Engines

      Windmills   Vertical axis
                                Horizontal axis

Electrical-Mechanical

      Dynamo/alternator
      Motor

*Can be constructed using a water piston

Figure 5.6.

Power Range of Energy Converters

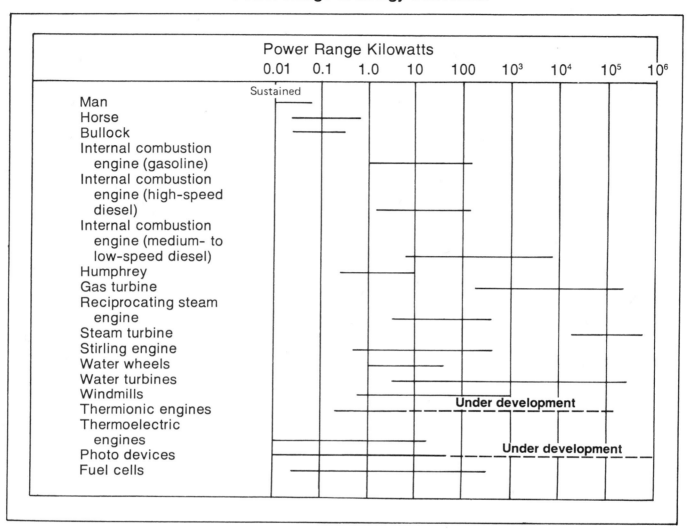

Figure 5.7.

hundred kilowatts; the steam turbine goes up to very high powers; and Stirling engines are similar to the diesel. Waterwheels go up to tens of kilowatts, water turbines up to megawatts, windmills could go up to megawatts. Thermionic engines, and thermoelectric engines, are not really worth bothering about, but I put them in for completeness. Photo devices, although they have been talked about in the multimegawatt range, I assume would give 10 kilowatts, and fuel cells will give something like 100 kilowatts.

In the developing countries we are concerned with 10 kilowatts and below, which eliminates the steam turbine and the gas turbine. So we have a fair number of possibilities that we can look at; our spectrum is quite large.

I think I have probably said enough now about the general picture. What I have tried to do is to indicate what the current world energy use is, how this might be effected by increasing use in developing countries, what the sources of power are in developing countries and the methods of conversion and the range which is suitable for use in rural areas of developing countries. I should like, now, to become rather more specific and look at some things that have actually been developed. I will outline one or two things that I happen to be currently associated with to illustrate the intermediate technology approach in this energy field.

Wind Energy

I mentioned renewable energy sources of which, of course, wind is one. Wind power is a very interesting possibility. It is free. Unfortunately, it is of very low density, although this does not matter much if you do not want much power, and we tend not to in any one place. It is variable, which can be a disadvantage but not if one is pumping water or grinding corn.

For the last 1,000 years the field has been dominated by the horizontal axis machine of the type pioneered in Holland. High-speed two-bladed mills are suitable for electricity generation, while the low-speed multi-bladed types are used for mechanical power—for grinding and pumping. Our contribution, I feel, is merely making available drawings of conventional designs and sending them out to developing countries for manufacture and erection.

There have, over the last 100 years or so, been developments in vertical axis machines—Flettner and Savonius rotors. However, for the design of the Savonius rotor you just take an oil drum and cut it in half vertically; then you mount the two halves on a vertical axis and the thing goes round. It has a good starting torque, low speed, and is suitable for pumping (photo 5.1). The great advantage is that you do not have to orientate it with the wind. We have recently installed two of those little units in Zambia for raising drinking water in a

Photo 5.1. Savonius rotor on test at Reading, England.

village community. Anyone can make it; we used wood when we were in Zambia.

There is need for much more work in wind power. One very interesting development now being pursued in Canada is a vertical axis machine with flexible blades. The great problem with all windmills is their great weight, and hence their cost. This thing has a weight of something like one-tenth of the weight of an equivalent horizontal axis mill. There is plenty of scope for good ideas, as well as conventional ones, in a simple field like wind power.

Muscle Power

For many applications muscle power is often the best solution. It is cheap, it is readily available, and it is often educational. For example, I was approached by a European space agency which is supplying television sets to rural parts of Africa. These sets require 35 watts of power for one hour a day. Now they have so far been using solar cells, battery packs, and small gasoline engines. Gasoline engines are unreliable and difficult to maintain, battery packs are very expensive, and solar cells are prohibitively expensive. Now this is a good example of where pedal power is what is required. A person can pedal very easily on a bicycle for an hour and generate 35 watts. We have in fact developed a generator of this sort. It consists of a small wooden frame with a wheel-mounted dynamo. I said it is educational, because if it goes wrong the reason is obvious; the dynamo has fallen off, or something similar. The local people do not regard it as a mysterious black box. They see what is wrong and they can repair it themselves and make more units of the same type.

Incidentally while talking about muscle power, its efficiency is interesting. If you assume that a man eats 2,500 to 3,000 calories a day (these are kilocalories), this is equivalent to about an eighth of a ton of coal a year. If you take our earlier assumption of working eight hours a day for 200 days a year at a tenth of a horsepower, it means that a man treated as an engine is about 10 to 15 percent efficient. It is really very good; it is almost as good as a petrol engine and costs far less.

Methane

The anaerobic fermentation of animal and vegetable waste is a good way of making fuel. Either cow dung or vegetation or both can be used as a starting material. And the resulting biogas will contain something like 60 percent methane and 40 percent carbon dioxide and a combination of hydrogen and other gases. This mixture has a calorific value of about 550 to 650 Btu.'s per cubic foot (about 550×30 joules per cubic meter). This is just about the calorific value of town gas. And it can be used for the same purposes. It can be used for lighting, heating, and for driving engines. Figure 5.8 shows a schematic of

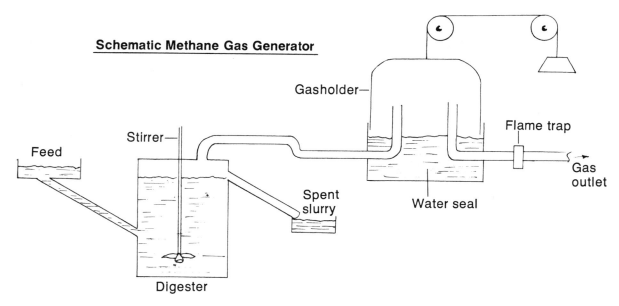

Schematic Methane Gas Generator

Feed

Stirrer—

Gasholder—

Flame trap

Gas outlet

Spent slurry

Water seal

Digester

Figure 5.8.

the sort of thing that we have been building. The main components are a digester and a simple gasholder. This can be based on the ubiquitous oil drum just upended and floating in water from which you take out your gas at a few inches of pressure. (One warning: you will notice where the illustration says: "flame trap." This is really very important. It is most embarrassing to have your methane plant explode. This happened to one of my research students recently and his friends would not speak to him for a week; we work exclusively with cow dung!)

Actually, cow dung is a source of energy that we should be considering in the developed countries. For example, in Great Britain we have the same number of cows as cars, so it is easy to work out how much dung you have. We have in fact, 1×10^8 tons of cow dung a year in this country. If you ferment all that, it is equivalent to about 3 million tons of oil. Therefore, cow dung is not an insignificant source of fuel even in the developed countries. We have recently developed a plant for 120 cows for use in the United Kingdom.

Liquid Piston Engines

Historically, the first engine to have a practical value, the Savery engine, and its development the Pulsometer pump, employed liquid pistons. These devices were superseded by the invention of the solid-piston beam engine, which enabled mechanical shaft power to be produced, and subsequent piston engine development has been almost entirely with solid pistons.

The liquid piston engine has a number of interesting characteristics, and, I believe, could play an important role, particularly in irrigation applications in the developing countries. Liquid piston engines are simple to construct, do not have tight tolerances, do not require lubrication of the piston, and have long life and low capital cost. The efficiency of these engines can be as high as that of their solid piston equivalent. Liquid piston engines can be constructed to operate with either external combustion or internal combustion. The simplicity of construction of liquid piston engines enables them to be manufactured on a small scale. This, together with the ease of servicing, is an important advantage when used in developing countries.

Liquid piston engines can be either—

1. external combustion (heat engines) which may be operated from any high temperature source, including solar heat, or

2. internal combustion, employing either liquid or gaseous fuels.

External Combustion (Heat Engines)

Solids, liquids, vapors, gases, and electron gases have all been used as the working substance in heat engines, and any nondissipative temperature-dependent property may, in principle, be used. Most development work with liquid piston devices has employed either vapors or gases as the working fluid. These engines operate either on a Rankine or modified Rankine cycle.

Figure 5.9 shows a number of engine configurations; the solar-heated Savery engine has been considered by J. R. Jennes. The basic engine can be considerably improved by reducing thermal irreversibilities; for example, the use of the insulating free piston in figure 5.9(c) to separate

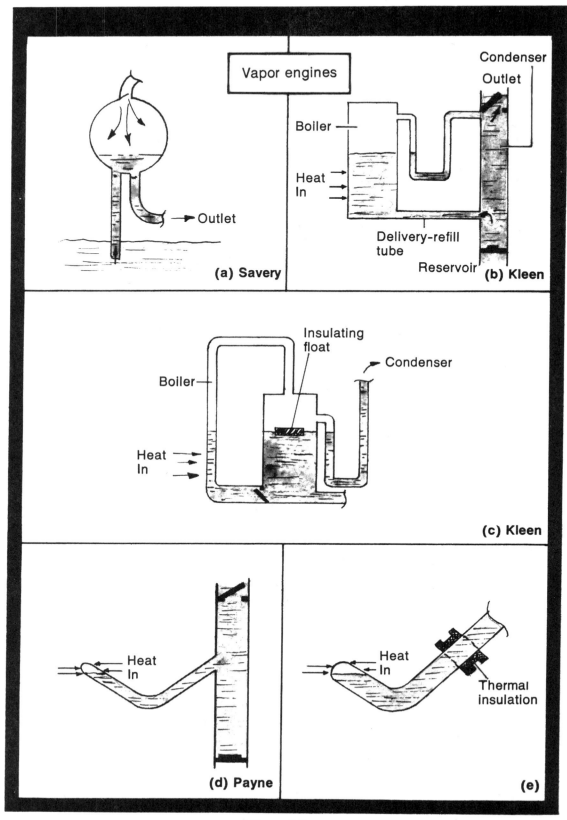

Figure 5.9.

the vapor phase from the cooler liquid and the addition of a separate boiler which also considerably reduces the volume of cold water added during the condensing stroke. We have found that the Payne configuration is considerably improved both in efficiency and operation by the addition of a thermal insulator as in figure 5.9(d) and (e).

We have been investigating some of these engines with a view to assessing their suitability for use with solar energy. The sort of device we are considering is the Kleen engine shown in figure 5.9(b). The pipe that goes down to the water you want to pump has a nonreturn valve and another pipe, again with a nonreturn valve, goes up to the outlet. If the device is full of water, as shown, you can apply heat to the boiler section and raise steam. The steam pressure will push down the water level in the U-tube forcing water through the outlet. At some point the steam will escape through the U-tube and become condensed. The resulting suction will draw in more water and the process can be repeated.

The thermodynamicists among you will say that such a process is not very efficient. It is not, but I have not shown it with all the regenerators and things; it would be too complicated, but we have not overlooked them. The engine is about one percent efficient as I have shown it, but we hope to raise the efficiency quite considerably. For this engine we will probably have to use a solar concentrator to get the temperature up. So far we have only built a small laboratory model which pumps a few cubic centimeters a minute; we have not scaled it up yet. It is still a new development.

Most of the work on this class of heat engine has been concentrated on the Stirling engine. This is a gas-cycle engine, as opposed to the vapor-cycle engines we have just been considering, and is usually used in conjunction with a solid piston, al-though liquid piston configurations, such as those shown in figure 5.10, are both feasible and attractive. It is a simple engine—a heat engine with diesel-type characteristics, though it will probably cost more than the diesel. It appears to have advantages for developing countries because it is a sealed unit and has no valves which should mean that it would require little maintenance.

Free-piston Stirling engines are based on the idea first suggested by Beale—figure 5.10(g). To operate on the Stirling cycle the power piston must lag the displacer by around 90°C. This is achieved in the Beale engine by arranging for a mass difference of 10:1 between the power piston and the displacer. The buffer space acts as an energy store to return the displacer piston at the end of the cycle. Figure 5.10(h) shows the application of the Beale principle to a liquid power piston. This concept has been extended by C. West in the Fluidyne engine in which the solid displacer piston is also replaced by a liquid. Three different methods are used to operate the displacer in the correct phase relationship to the power piston. These are a rocking beam, a pressure feedback, or a jet stream, illustrated in figure 5.10(i).

Internal Combustion

The Humphrey engine has probably reached a more advanced stage of development than other liquid piston engines. It operates on the Atkinson cycle, and a general diagram and details of the valve operation are shown in figure 5.11. At the beginning of this century H. A. Humphrey constructed several large pumping engines fuelled by gas which operated satisfactorily for many years. For various reasons, however, interest was lost and development discontinued. In 1970, however, work was restarted on these pumps in my laboratory by R. J. Congdon and continued

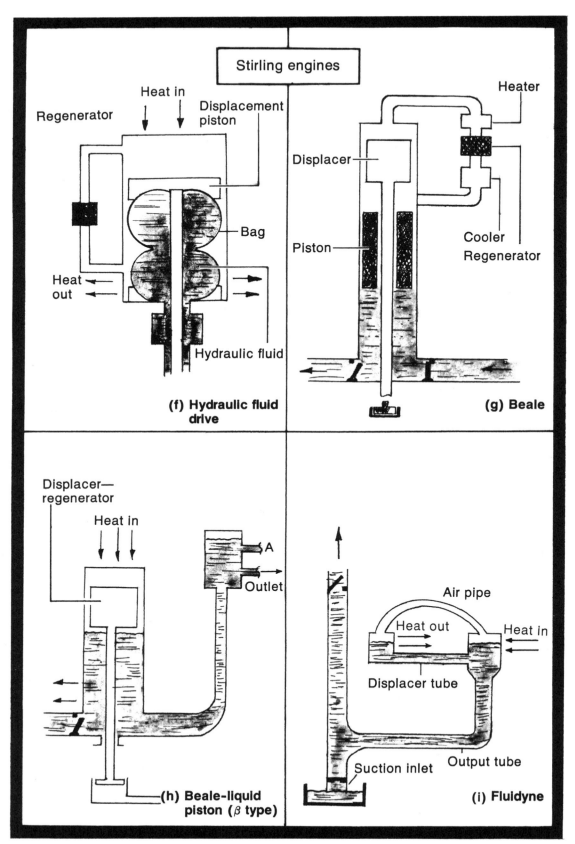

Stirling engines

Regenerator

Heat in

Displacement piston

Bag

Heat out

Hydraulic fluid

(f) Hydraulic fluid drive

Heater

Displacer

Piston

Cooler
Regenerator

(g) Beale

Displacer—regenerator

Heat in

A

Outlet

(h) Beale-liquid piston (β type)

Air pipe

Heat out

Heat in

Displacer tube

Suction inlet

Output tube

(i) Fluidyne

Figure 5.10.

76

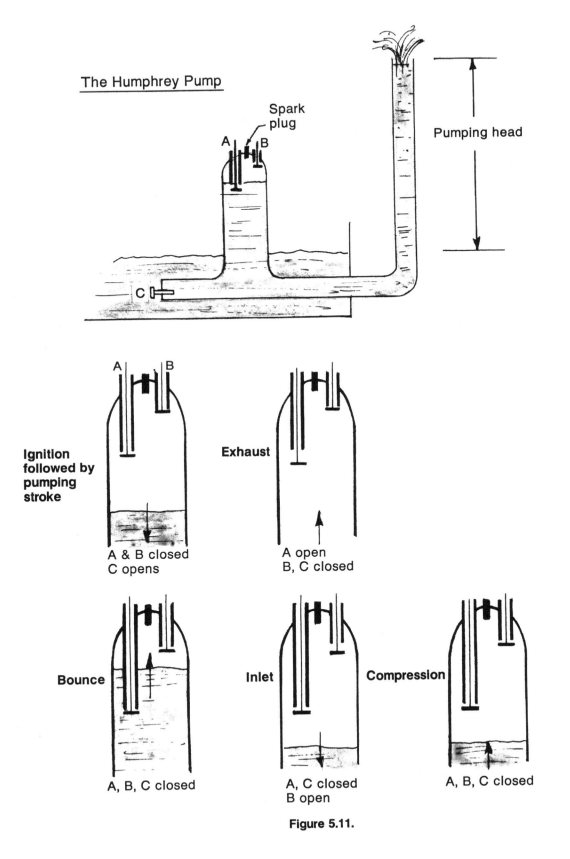

The Humphrey Pump

Spark plug

A   B

Pumping head

Ignition followed by pumping stroke

A & B closed
C opens

Exhaust

A open
B, C closed

Bounce

A, B, C closed

Inlet

A, C closed
B open

Compression

A, B, C closed

Figure 5.11.

to the point where we now know quite a lot about their principles and operation.

The layout of the Humphrey pump is as shown in the illustration. If the device is full of water as indicated and all the valves are closed, the gas in the cylinder will be compressed. If this is an explosive mixture of gas which you ignite with a spark plug as in a car, it blows a column of water down the cylinder and up a 10–20 meter head. At the same time the water valve opens and takes in more water; then the column returns. The exhaust opens and the spent gases pass through it. Since it is lower than the rest of the cylinder head, you are left with a closed gas space which is compressed. At this point the inlet valve opens, takes in a new charge, the column goes up, again comes down, compresses the new charge, and the process is repeated.

Photo 5.2 shows the latest six-inch-diameter version which pumps water from the tank in which it is immersed. Photo 5.3 gives a close-up of the cylinder head and shows the rubber bag which is used as a gas supply reservoir, the valve stems and their springs, and on the right, a piston device which produces a spark via a conventional coil at maximum pressure.

The Humphrey pump is a most attractive solution for water pumping. First of all it has a high efficiency, which is comparable with that of a diesel engine driving a pump. Secondly, it is clearly very cheap because it is made out of pipe and has no tolerances and a very simple head. Thirdly, it is suitable for local manufacture.

Fourthly, most important, it is readily maintained because if you make it locally, you can maintain it locally. Low-head pumping is a very important requirement in many developing countries where one wishes to move water out of irrigation canals, into things like paddy fields. So this pump, we feel, is the sort of thing that could well be developed in Europe, but made and introduced in developing countries.

The first four-inch pump we built only had an efficiency of 1.5 percent, but we have steadily improved this until our latest pumps have an overall efficiency of about 25 percent and will pump about 3,000 gallons (13,500 liters) per hour. The next phase of the program is the modification of the design to operate on liquid fuels and to develop an ignition system which does not require a secondary battery.

Energy is the great common denominator of both developed and developing countries, a fact which has been underscored by the permanent energy shortage. Therefore, we must explore all the available intermediate technologies in the energy field: windmills, water power, solar stills, methane generators, unconventional engines (Humphrey pumps, solar engines, and Stirling engines), and conventional engines (including diesel). Furthermore, we should give renewed consideration to human power, which is adequate for scores of everyday tasks. It is in the interest of all countries to design, perfect, and employ low-energy alternatives.

Photo 5.2. Humphrey pump in supply tank.

Photo 5.3. Humphrey pump on cylinder head valves, ignition, and gas reservoir.

Chapter Six

Pedal Power

by Stuart S. Wilson

I consider the bicycle as the most important modern invention. It represents the breakthrough in modern technology. I will just give a few illustrations to show the salient features of this remarkable achievement. It took about 5,000 years, of course, from the wheel to the bicycle, and the bicycle itself took another 50 years to evolve. Having evolved it has hardly changed, but it was a remarkable evolution.

Evolution of the Bicycle

Photo 6.1 shows what I regard as the first "proper" bicycle which was made in 1839 by a blacksmith in Scotland. It is a nice bit of ironmongery, but it is hardly lightweight nor is it yet an efficient piece of technology. But it worked. It was a technical success and a commercial failure.

The Ariel of 1870 (photo 6.2) represented a considerable technical advance, having a lightweight spoked wheel. It was still crude in the sense of needing torque arms and tie bars to transmit the torque from the hub to the rim. But it was very advanced in having a large diameter wheel, which is very efficient in minimizing rolling resistance on either rough or soft ground. In England we call it a "Penny Farthing" or an "ordinary" to distinguish it from the later "safety" design; in America, they call it a "high wheeler." We still specify the gear ratio of a bicycle as it was done for this sort of machine; it is still expressed as an equivalent diameter of a

Photo 6.1. MacMillan's bicycle, 1839.

directly pedalled wheel. So the normal bicycle with a 26-inch (0.66 meter) wheel and 46 teeth on the front sprocket and 18 teeth on the back is equivalent to a wheel of a 66.5-inch diameter—roughly 1.7 meters. The gearing of the "ordinary" is clearly limited by the length of the rider's legs. It was very efficient and lightweight, particularly the tubular construction. But the structural engineers of that time just did not realize how good their construction was; they did not start using it for bridges and other things for another 30 years.

The Rover safety bicycle of 1885 (photo 6.3) is the prototype of the modern bicycle. The year 1885 was a very crucial year because it marked the definitive form of the bicycle and the beginning of the motorcar. Gottlieb Daimler and Karl Benz produced their first vehicles in 1885 based, of course, firmly on bicycle technology. This design had chain drive to get the best gear ratio and this led to the evolution of a very efficient chain—the bush roller chain—by Renold, a Swiss who worked in

Photo 6.2. The Ariel "Ordinary," "High Wheeler," or "Penny Farthing."

Manchester, which in turn led to the establishment of a large industry. Ball bearings were developed for the wheels and pedals and this led to the foundation of the ball bearing industry. The big wheel used

Photo 6.3. Rover safety bicycle, 1885.

before this machine did not really need a soft tire. But once the wheel was reduced to this size a pneumatic tire was needed and was indeed reinvented in 1888 for this purpose. By 1888/90 the tire had become a technical and commercial success that led to the establishment of Dunlop and all the other tire firms. The large electrical firm of Lucas, again, owes its foundation and commercial success first of all to oil lamps and later electric lamps for bicycles and then for cars. And most of the major car firms, such as Rover, Hillman, and Singer in Coventry, and then Morris in Oxford, were all originally bicycle firms.

So you can see why I maintain that the bicycle is the breakthrough in modern technology. All these big industries started with the technology and commercial success of the bicycle. Indeed, production engineering had a lot to do with the bicycle, and the bicycle had a lot to do with production engineering. Bicycles were being made by the million before Henry Ford made his first motorcar and certainly before he made cars in quantity.

Not surprisingly, we can follow the influences of the bicycle down to the present day, to an instance where man has used it to fly about a kilometer in a straight line. But long before that, the Wright brothers were bicycle makers. If you look at their Wright Flyer, you will see exactly how much it owes to the bicycle. In fact, all early airplanes were full of bicycle technology. Therefore, I think the claim for bicycles is justified.

Social Change

Technology, I believe, is the instrument of social change. And one can see this again with the bicycle. It was intimately connected with Women's Liberation; education for women and the bicycle were the great "freedom factors" of late nineteenth-century Europe. But perhaps

Photo 6.4. Pedal power used for flying.

the biggest social change that the bicycle brought about was due to its successor, the motorcar. In the last 50 years the motorcar has affected life in Europe and America beyond recognition. So technology is a great instrument of change, not necessarily progress, of course; it may not be a change for the better, but certainly an instrument of change.

Figure 6.1 shows the energy efficiency of various modes of transport. It is difficult, of course, even to define what we mean by energy efficiency; one definition is energy units required per unit of weight per unit of distance travelled. The graph is plotted as body weight in kilograms against all-up weight, not weight per passenger. Among the flyers, a heavy bird such as the pigeon has the greatest energy efficiency, which is why it can cover such long distances at high speed. As the flyers get smaller, they get less efficient to energy terms.

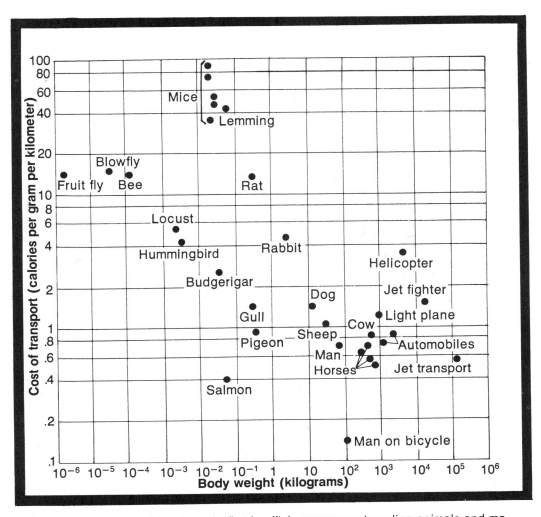

Figure 6.1. Man on a bicycle ranks first in efficiency among traveling animals and machines in terms of energy consumed in moving a certain distance as a function of body weight. The rate of energy consumption for a bicyclist (about .15 calorie per gram per kilometer) is approximately a fifth of that for an unaided walking man (about .75 calorie per gram per kilometer). With the exception of the black point representing the bicyclist (*lower right*), this graph is based on data originally compiled by Vance A. Tucker of Duke University.

Similarly, large jet transports are most energy efficient and helicopters least. It appears to be a case of bigger is better; but of course this is only one aspect—energy consumption. There are lots of disadvantages to large things. But the two exceptions to this general rule appear to be young salmon, and I suspect other fish and dolphins too, and the man on the bicycle as compared to a man walking or running. The high energy efficiency of the bicycle is based partly on the efficiency of the large-diameter wheel, the pneumatic tire, and ball bearings and partly on the difference between mechanical work and what one might call physiological work. If I press down hard on something and exert a force, but there is no movement, I am clearly not doing any mechanical work but I am using up muscular energy in exerting the force. If I am standing, I am using muscular energy just to keep standing because the muscles, in tension, and the bones, in compression, must act together to keep me up. If you imagine that the heap of bones of a skeleton was erected you would have to provide a large number of tensile forces to keep it in equilibrium. So when I stand I

Photo 6.5. Bicycle fitted with a dynamometer.

use more energy than when I sit, and that uses more energy than when I lie down. So there is this isometric or isotonic work which is needed when one is walking, but on the bicycle one is sitting down and less of this work is needed.

Now the other sort of work is what I call "shadowboxing" work—movement but no force, where no mechanical work is done. For example, if you are walking everything is swinging and you are using up this shadowboxing work, but on a bicycle most of your body is stationary. Admittedly, the upper part of the legs are moving, but even the feet are going around at constant speed, so you minimize this shadowboxing work. Putting these various things together I think you can begin to see

why the bicycle is so efficient in energy terms.

Very simply, it is a matter of using the right muscles (the leg muscles which are the strongest in the body), in the right motion (the rotary pedalling motion), at the right speed, then transmitting the motion efficiently (through ball bearings and the roller chain) and then of using it efficiently with a pneumatic tire. One must also minimize the weight and the wind resistance as far as one can, although that is the most inefficient part of a bicycle as you well realize in a head wind.

Photo 6.5 shows a bicycle that we fitted with a dynamometer so that we could actually measure the power needed to drive the machine. The crankshaft is

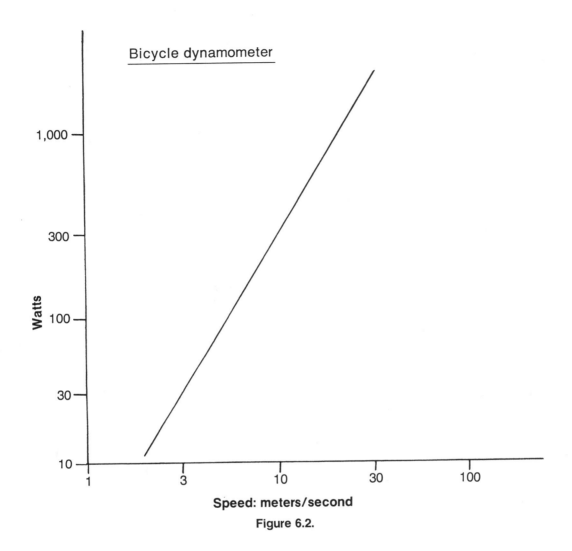

Figure 6.2.

separate from the sprocket and the torque is transmitted through an arm which has wire-resistance strain gauges. The strain registered by them is picked up by a zigzag wire to give a signal proportional to speed. A radio transmitter and battery transmit the signals by telemetry to the frame where they are multiplied together. One can ride along and read off the road speed, the horsepower, and the torque on the instrument panel in front of the handlebars. A typical plot of power against speed is shown in figure 6.2. The normal cyclist would have an expenditure of about 75 watts—roughly one-tenth of a horsepower. I regard 75 watts as the fullest sustainable output of the human body, using the right muscles, right motions, and the right speed. Of course, one can do more than that for shorter periods, and in fact most cyclists can do up to 750 watts for a few seconds. If one is a weightlifter, or perhaps a docker handling a large piece of cargo, instantaneously one may be working at three times that amount. It is a very valuable feature of the human body that one

Photo 6.6. Chinese pedal cart of elderly design.

Photo 6.7. Taking the pigs to market (Vietnam).

Photo 6.8. Rickshaw in Dacca, Bangladesh.

Photo 6.9. Mark I cycle rickshaw.

Photo 6.10. The simple differential gear.

can work with 10 times the normal output for short periods. There are not many engines or motors that will do that.

Traditional Rickshaws

What other ways are there of using this power transmission system, apart from the bicycle? The Chinese pedal cart is one way of doing it. It has some rather nice Gothic welding and doesn't bother with chain tensioners. It is very practical though not, I somehow feel, optimized. Photo 6.7 shows one way of getting three pigs to market in Vietnam. The layout with two wheels in front has the disadvantage that the rider has to be perched rather high up to see over the hood when it is up. And also there is nothing very subtle in the way of steering. The whole body is pivoted about the central point and is very heavy to control; if one of the wheels hits a bump or a pothole, it will cause a terrible wrench on the whole system, though it does mean that one can use the normal drive on the back wheel.

Photo 6.8 is a cycle rickshaw as used in Dacca, in Bangladesh; it is beautifully painted but is very crude mechanically. This rickshaw has a solid sprocket with no freewheel. There are no bearings in the middle of the rear shaft, only where the wheels are—and I don't think those are ball bearings. So the rear shaft has to be very heavy and solid to take the pull of the chain, which is roughly twice a man's weight. The cycle rickshaw uses the gear ratio which has been optimized for normal bicycles but which is quite wrong for something that might weigh three times as much with its passengers and load. This design originated about 1890 in Britain and went to India and has just been copied ever since. It has never evolved. I think the potential of pedal power is such that one should try and evolve it for other purposes.

For example, a rickshaw without gearing is very difficult to start even on the flat; with a load it is very difficult to get up anything of a gradient.

Improved Rickshaws

My first attempt in 1973 at making a cycle rickshaw was designed in a week and made in another week. Its only technical virtue is in its back axle. If most of the weight is on the back wheels then, for good traction, one should attempt to drive both wheels. But if you have a solid axle, it is very difficult turning corners. So one needs a differential, as on a normal motor vehicle, but they are expensive, so I evolved a very simple arrangement (photo 6.10). There are two half-shafts using bicycle-type bearings. On the left-hand

Photo 6.11. Neo-Chinese sailing wheelbarrow on show at Oxford, England.

Photo 6.12. Hand-powered machine for rolling channel sections in making large wheels.

shaft there is a normal sprocket and freewheel; on the right hand one there is another sprocket and freewheel except that this has 24 teeth instead of 18, to give a lower ratio, and the two are connected with pins. Normally both wheels are driven, but on a corner the inner wheel, whichever one it is, takes the drive and the outer wheel, which is trying to go faster, freewheels, and so it gets around the corner—not as well as the normal differential, but it gets around. In slippery conditions, whether due to mud, loose gravel, or even ice, this design achieves the same effect as a limited-slip differential, because if one wheel slips then the other wheel takes the drive. So it is a simple solution to the basic problem of traction.

Chinese Wheelbarrows

There are other uses for bicycle technology such as the updated version of the traditional Chinese wheelbarrow. The Chinese invented the wheelbarrow thousands of years ago and very sensibly they, firstly, had a large-diameter wheel; secondly, they put that wheel under the load, or nearly under the load because you still need some weight on the handles. Like a trailer behind a car you need some weight on the bumper, but not too much. Thirdly, they sometimes used a sail when the wind was favorable. This design is updated merely by having a bicycle wheel with a pneumatic tire instead of the traditional wheel. But it still retains the load largely over the wheel.

In Oxford we made a neo-Chinese sailing wheelbarrow (photo 6.11) for a Manchester clergyman who wanted to walk 2,000 miles across the Sahara—with the wind, in order to raise money for charity. He wrote and asked if I could make him a two-wheeled trolley to carry a load of 350 pounds. I thought two wheels was a bad idea because on rough ground one wheel would hit a bump, a hollow, or a soft patch at a different time from the other wheel and would tend to slow the whole cart. It would do this continually, so he would be fighting it for 2,000 miles across the Sahara. I suggested the Chinese wheelbarrow with the large-diameter wheel would be better.

The man from Pakistan who constructed it told me of a very ingenious technique for making a solid rubber tire for wooden wheels. The inventor made a steel channel which was "shrunk" onto the wooden wheel in the traditional way. Then he cut strips of old motorcar tires which he clamped in the channel.

To make a steel channel, to curve it round into a wheel, and to get it the right length is quite a skilled piece of manufacture. No doubt it can be done easily in Pakistan, but it is much too difficult to do in Oxford University.

So I designed and made a very simple rolling machine with a flanged roller which can be moved up and down with a screw and two plain rollers and a pedal used as a handle (photo 6.12). It can

Photo 6.13. Oxtrike chassis.

Photo 6.14. Rotary pump fitted with pedals.

transform straight channel into curved channel. Incidentally, even the bearings are bicycle-type, bottom-bracket bearings, and the whole machine could be made anywhere and has, I think, a lot of uses. For the construction of this big wheel I suggested a spoked wheel, but it is difficult to get spokes of sufficient length, so in fact we took pieces of foamed polystyrene, used for insulation, about 25 millimeters thick. Then we glued hardboard 3 millimeters thick either side, which formed a sandwich construction. This is a very efficient form of construction much used in nature. Our skulls are of sandwich construction—a thin layer of bone either side of a cellular construction (an almond nut is difficult to break because the shell is a sandwich construction). Then we had a wooden rim, made from 12 overlapped pieces of wood. We protected this with a steel channel which had the opening on

the outside. By interchanging the rollers we were able to roll it in and put it on.

The platform also is made from sandwich construction and the sail is square-rigged about two meters by one, which evidently was enough to help. The wheel is about 1.2 meters diameter and it is only 50 millimeters across the tire, but it seems to perform quite well in sand. So I think this principle of the Chinese wheelbarrow with the large central wheel is a very good principle, and I think it could be used much more widely than it is. The European wheelbarrow, for some reason, has a small wheel stuck in the front instead of a big wheel in the middle.

The Reverend Geoffrey Howard succeeded in crossing the Sahara on foot, 2,000 miles from north to south. He averaged over 20 miles per day, despite many difficulties. It was a personal triumph, but should also serve to show the potential of the Chinese type of wheelbarrow.

Cycle Production

Oxfam, the international charity based in Oxford, is convinced of the whole philosophy of intermediate technology and in particular is convinced of the potential of pedal power and muscle power, because this is one way by which people can help themselves if they can be motivated to do so. They have been providing financial aid for my work on improved cycle rickshaws. As I have mentioned, the first cycle rickshaw was made in a week, and was fairly crude in design. Normally, however, bicycles are sophisticated, requiring, for instance, steel tubes of a special alloy. I understand that India has to import most of the steel tubes which she uses to make her two million bicycles a year. But on the other hand, India does have steel mills which can make sheet steel. A design to use standard bicycle parts and sheet steel which we call the Ox-

Performance of some simple water-lifting devices					
	Number of:—		H.P. equiv. of water pumped	Horsepower per:—	
	Men	Bullocks		Man	Animal
Picottah	4	2	0.10	0.025	—
Wellsweep	5		0.20	0.04	—
Mhote	2	2	0.23	—	0.115
Persian wheel	2	2	0.30	—	0.15
Inertia pump	1		0.08 to 0.12	0.08 to 0.12	—
Pedal-driven pump	1		0.1 +	0.1 +	
Horsegear			0.3		0.3

Figure 6.3.

trike is shown in photo 6.13. It is almost entirely made from sheet steel of 1.6-millimeter thickness, because that is a standard size throughout the world and can be welded, brazed, rivetted, or even bonded with epoxy glues and self-trapping screws. The basis is a box-section girder running along the tricycle's length. The girder is made from sheet steel which is cut with a foot-operated guillotine and folded with a hand-folding machine. Then another channel section is put inside it to form a square section 63 millimeters wide and 75 millimeters deep. The two are then welded or spotwelded or brazed or rivetted or joined by any other means. I think this construction could be made anywhere in the world.

The brakes on the traditional rickshaw design are very crude; the difficulty is that with the wheels stuck out, there is nowhere to put conventional brakes. There is no fork to mount them on and drum brakes would be a little sophisticated. In answer to this design challenge we used inboard band brakes. They are operated by a foot pedal and are very powerful, and

there is a handbrake operating on the same pull rod. The gearing of traditional designs is another weak point; we wanted to introduce three speeds, which may sound a luxury, but after all every vehicle with an engine has at least a three-speed gearbox, and it will greatly extend the utility if you can have a very low gear for starting. The normal bicycle has a gear of 66½ inch, but this one has a bottom gear of 31½ inch, which is less than half; the middle gear is 42, and the top one is 56 inch. So even the top gear is lower than the normal one and this is about right; it makes a big difference to riding it to have this low gear. We use a standard Sturmey-Archer, three-speed hub gear as an intermediate gearbox, as is used on a motorcycle. The primary chain drives a detachable sprocket that you can change (there is a choice of 16, 18, 20, or 22 teeth). Then there is a sprocket rivetted to either side of the gearbox, and two chains. Although I am not sure that it is an optimal solution, it works.

The machine is fitted with a small truck body and the payload is about 150 kilograms. The back of the body is short so

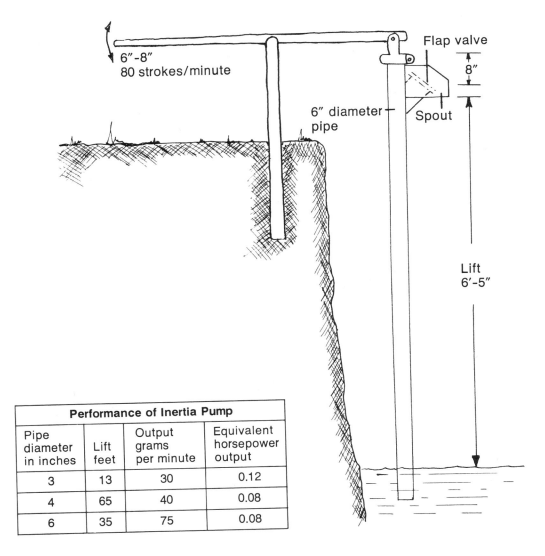

Performance of Inertia Pump			
Pipe diameter in inches	Lift feet	Output grams per minute	Equivalent horsepower output
3	13	30	0.12
4	65	40	0.08
6	35	75	0.08

Figure 6.4.

that the vehicle can be tipped up on end, which is very useful for servicing or for parking or for tipping out loads like sand or gravel. The main frame weight is about 13 kilograms.

Stationary Pedal Power

So far we have discussed transport uses of the bicycle, but clearly there are a lot of stationary uses. The bicycle has been optimized for one purpose and I don't think it is an optimum for stationary purposes (for one thing you want to take the drive forwards so you can see what you are doing, rather than backwards). There are three approaches to adapting it to stationary use: you can modify a normal bicycle, or you can modify something that is hand-driven by putting pedals on it, or you can design an optimum stationary pedal-power unit.

Photo 6.15. Chinese square-pallet chain pump.

Photo 6.16. A one-man dynapod in Uganda.

Photo 6.17. A two-man dynapod driving a corn grinder. (A winnowing machine is in the background.)

Photo 6.19. Rodale winch.

Photo 6.18. Rodale Energy Cycle—churning butter.

95

The rotary pump shown in photo 6.14 with one end cover removed is normally hand-driven but has been fitted with pedals. The pump itself is an eccentric-vane type. It has plastic vanes, sliding in slots, and there are springs on rods pushing each pair of vanes apart. There is only a small variation of length as they go round. It is a very simple design and I think it could be produced cheaply in quantity, although it isn't at the moment. It is such a simple type of pump and suitable for direct pedalling for heads of say three to eight meters that it is, I think, worth developing, perhaps by altering the materials of construction.

Some figures for the performance of some traditional water-lifting devices in India are given in figure 6.3. They show how four men only achieve 75 watts—one-tenth of a horsepower between them, and even two bullocks and two men at best only achieve 225 watts—0.3 horsepower. The figures mean that if one had some other form of power, perhaps electricity or diesel, one would only need a very small amount of power to replace such a device. But that may not be the right thing to do; it may well be better to improve the efficiency, to double or even treble it, and get more water for a given input. This was what really started me thinking on means by which existing designs could be improved and increases in energy efficiency could be attained.

Figure 6.4 shows one very simple form of pump. It is just a horizontal handle like a seesaw which suspends a pipe which has no foot valve, only a flap valve at the top. At first sight it looks too simple to work, but in fact it is due to the inertia of the column of water. When the pipe comes down, the water remains stationary and is delivered through the spout, and as the pipe goes up, the valve closes and the whole column of water is raised. It appears to work quite well—around about a tenth of a horsepower again. There is an even

simpler version in which you have just a pipe, and instead of a valve you have a small boy with his hand over the top of the pipe. This method contrasts with the traditional Chinese method of raising water, a square-pallet chain pump, which is shown in photo 6.15.

The third principle of adapting pedal power to stationary uses is to design an optimum stationary pedal-power unit, which I call the "dynapod." It employs a saddle, pedal, a chain, but it takes the drive forward. I theorized that you would be able to gear it down for something like a winch or gear it up for a winnowing fan.

Recently, a number of dynapod-like pedal units have been constructed. Alex Weir of Edinborough University made a version in Uganda. A more sophisticated unit called the Rodale Energy Cycle, which can perform numerous tasks around the home and homestead, is also a very versatile tool in the garden, where it has been used to cultivate, plow, and weed. In fact, a separate winch assembly, which has a test pull of 1,000 pounds, has also been developed by Rodale Resources. These pedal devices suggest that human power, particularly leg power, can be a genuine force in the home or on the farm. And significantly, these inventions indicate that developed and developing countries alike have something to gain from practical applications of intermediate technologies.

I believe that the more attention given to pedal power possibilities, the more practical applications will be discovered. In fact, it is fairly safe to assume that many machines with hand cranks could be easily converted to pedal drive. For example, a common winnowing machine (photo 6.22) has a chain drive and a pedal as a handle—but it would be much better to pedal it.

Interestingly, our nineteenth-century ancestors enjoyed a more sophisticated winnowing machine (which actually had its roots in fourteenth-century China). The operation is simple: a crude, four-bladed fan blows the air across the hopper; the fine dust is blown away. Then the sieves are shaken and the husks are loosened and blown off. To allow the grain to come down, there is a fine sieve which will, in turn, catch the heavy particles.

The winnower, then, is quite a sophisticated design that can be hand-driven, but would be much more efficient if pedalled. Therefore, I think this design needs rethinking for optimum effectiveness. Similarly, the small motor-powered rice thresher (photo 6.23) would benefit by such a rethinking and adapting. There have been attempts to pedal it, but it really needs two men and a flywheel to work well.

Significantly, there is no limit to simple machines that could be improved if pedal adaptations are made. For instance, figure 6.6 represents a sketch of a peanut thresher which originated in Malaysia. The drum can turn in only one direction, so you could not pedal it from the front; it would be going the wrong way. That is

why the thresher has an elaborate arrangement with a crankshaft and a treadle. Because one man has to stand on one leg in front, treadling with the other, he could not bend down to pick up the peanuts to feed the machine. He had to have an assistant to pick up the peanuts for him.

To eliminate some of this complexity, I suggested that it would be much better to have the assistant on a saddle at the back pedalling and driving the drum; then the other man could feed the machine from the front.

Occasionally, it is possible to replace a motor drive with a pedal unit, although indications are that this procedure could benefit from some fundamental research.

Perhaps the most encouraging news is that pedal power inventions, such as the cassava grinder from Nigeria, with hacksaw blades set in a bicycle wheel (figure 6.7), are coming to the forefront by the day. These developments should not only give all of us cheer but should suggest that pedal power is a subject worthy of serious scientific scrutiny.

Photo 6.20. Rodale winch with cultivator.

Photo 6.21. Rodale winch with tractor utensils.

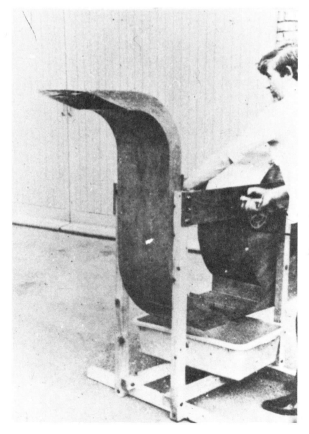

Photo 6.22. Hand-driven winnowing machine.

Photo 6.23. A small motor-powered rice thresher.

Photo 6.24. Ransomes/Hunt pedal-driven maize sheller.

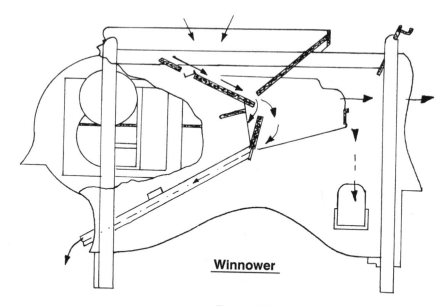

Winnower

Figure 6.5.

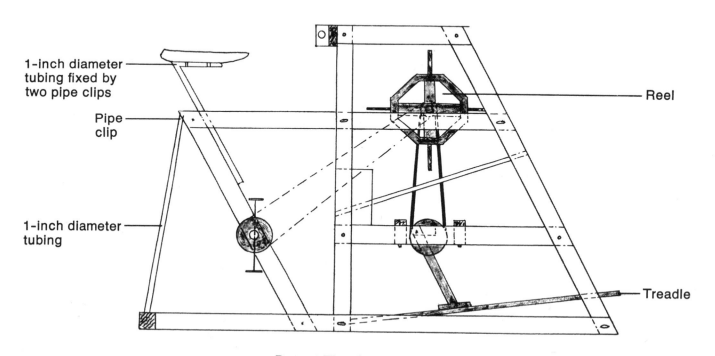

1-inch diameter
tubing fixed by
two pipe clips

Pipe
clip

1-inch diameter
tubing

Reel

Treadle

Peanut Thresher with Pedal Drive

Figure 6.6.

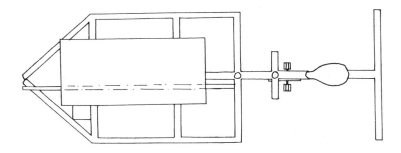

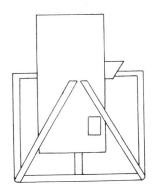
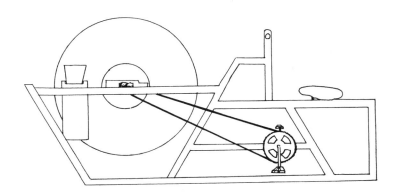

Figure 6.7. Cassava grinder from Nigeria.

Chapter Seven

Intermediate Chemical Technology

by George F. Reynolds

The major contrast between high technology and intermediate technology is the complete reversal of the general trend towards bigness, in other words the emphasis is on labor-intensive rather than on capital-intensive systems. It is generally accepted that the source and the center of world poverty are in the rural areas and so the principle hope and objectives of intermediate chemical technology, as indeed of all intermediate technology, are to help the poor in developing countries to work themselves out of their poverty by providing them with knowledge and training and details of processes at a level that they can understand, together with the right sort of equipment, in the right place, and at the right price. It is a desirable feature, not always easy to achieve, that the intermediate-technology process plant, once it has been established, should be capable of expanding in scale so that, once they have gotten used to producing something on a small scale, they can move to a somewhat larger scale.

All these general concepts apply to chemical technology and chemical technology occupies an important place in the general framework of rural development. Such technology offers particular opportunities for employing many more local resources to raise the standards of living and health, while it diminishes dependence on outside sources. The processes themselves are often those which have been discarded or superseded in more "developed" countries on the grounds that they require too much labor, or that they cannot be operated continuously. Those three reasons why

they have been discarded in developed countries are three very good reasons why they should be considered in developing countries. Although in any particular area available materials and local conditions determine what can be done, in general the objective should first be to produce basic materials or chemical compounds and then either to use these in simple manufacturing processes for local benefit or in simple processes to provide the community with something which it can sell in order to obtain money.

Locally Added Value

A very important part of intermediate chemical technology in villages and rural communities is that the people should be able to take local raw materials and partly process them. In other words, they should be able to produce a crude product which can then be sent somewhere for further refining. This is a very important part of intermediate technology. There are so many biological materials, local products, which are of considerable value not only to the country at large, but also internationally, that are never exploited because the transport situation is such that you cannot move many tons of a natural product, such as a plant, over a virtually roadless country for processing in the towns. For one thing, the material, if it is biological, has probably deteriorated by the time it reaches the factory in the town. But by using intermediate-technology processes in the village, several tons of a plant can be reduced to perhaps two or three liters of a highly valuable essential oil, which can then be transported easily and sold for a high price in the town. The oil is not pure, but it has become highly valuable. A pharmaceutical company in the town can then process it further and immediately the village community has a source of income and the urban communities have a source of raw materials which was formerly closed to them.

Batch Processes

A number of things have to be taken into account, when considering what process to start. First, the process should be simple and it must involve a minimum of equipment, locally made, if possible. Power requirements should be labor intensive and it should be batch process; continuous processes are far too complicated. It is much better to work in batches which should be operable on a scale which ranges from 20-liter oil drums up to 4,000 or 5,000 liters. The large oil drum, which seems to be the major vessel in all developing countries, is a very useful basis for process equipment and can be turned into all sorts of very useful pieces of process plants, as we shall explore later. All these criteria, of course, are completely opposed to those of modern industry and it is therefore often very difficult to scale down the modern processes for use in intermediate technology. But as I have already mentioned, details of suitable processes do exist, because they parallel the processes which were used before the industrialization of Western countries. For intermediate technology, then one consults old books, the histories of chemistry, and one looks up the historical background of modern industrial processes.

It is also surprising to find how many small-scale batch processes are still in profitable operation around the world, certainly in Great Britain and also in various parts of Europe. They remain a commercial proposition, because they have a specialized product, which is not required in large quantities and nevertheless is expensive and which has a limited market. For example, I helped to operate a very successful process for the extraction of specific drugs, including one for treating Parkinson's disease or palsy. These drugs were extracted from an African bean and virtually all the work was done in batches with the extractions being carried out in plastic dustbins. Then we simply had a 500-liter stainless steel tank, a 50-liter homemade filtration plant, and a 20-liter glass distillation apparatus; this was the entire equipment of the factory. All the drug for all the initial clinical trials was made by this method, and all the drug that was necessary for medical purposes in the early stages was supplied from this factory. Later on, the use of this drug became very widespread and then big pharmaceutical companies came in with continuous synthesis. The operation was no longer profitable for this small company which therefore moved on to do yet another new drug for which there was not yet the demand to attract the large pharmaceutical companies. So there is a surprising number of small processes still going on using virtually all intermediate-technology equipment.

The great attraction is in developing processes which involve natural products, such as this African bean which we used, because the materials are easily obtained and are generally available. There is generally something available locally, especially in tropical areas, which is useful, many are biologically fossil materials such as fuel oils.

Energy

An important criterion for processes in intermediate chemical technology is that the power requirements should be small. You have to assume that electricity and fuel oil will not exist in the communities; you have to think of other methods. The combustion of wood is a common sort of energy that can easily be utilized with the advantage, once again, that wood is biologically regenerated. If you do not use it too fast, you can grow more trees, although this is not always the pattern in developing countries. But, after all, the use of wood in this way is really very inefficient and it is a very wasteful means of supplying energy, both in relation to the energy that is produced for a given weight of fuel and because important chemical substances are lost. Wood is a very important and interesting chemical and you shouldn't just set light to it. You can convert your wood to charcoal which is done very often, and you will get a material which gives a much greater intensity of heat; but in making charcoal you lose the volatile products of the wood, which are very valuable. However, it is possible to develop processes for distilling wood in a type of retort before final conversion to charcoal and we will mention this a little later. It is also important to remember that charcoal is not only an important fuel, but it is also a valuable reducing agent when one is smelting metals from metal ores.

Another very important fuel which has not been adequately developed or exploited is methane, which can be generated from animal dung and from other waste products. You can get a high yield of methane if it is allowed to decay under the right conditions. The important feature of this is that the value of the raw material, the dung, as a fertilizer is enhanced rather than impaired, so that in fact, the village farmer does not lose anything in the process. In practice many developing countries dry and burn animal dung as a fuel whereupon its value as a fertilizer is lost and thus they are always desperately short of fertilizer. The dung from three cattle is sufficient to generate

methane, which has a calorific value equivalent to about four to five liters of diesel oil per day. Therefore, methane is obviously a very important source of fuel and a very important saving in money. It has also been realized that many other materials can be used for fermenting to produce methane, providing there is some dung present to supply an initial culture to the bacteria. And so you can feed into the plant a mixture of dung and vegetable waste and produce the necessary methane without necessarily having three cows per family.

The equipment for producing methane is very simple, it can be made locally. An average family, for its own domestic uses, requires about 2.5 to 3.0 cubic meters of methane per day. For this, one needs three cows, or one or two cows and a mixture of dung and organic waste material. One of the problems with producing methane is that the organism which carries out the digestion really requires a temperature of 30° to 35°C. for an optimum yield. This is not too serious a problem in hot countries and since the fermentation does produce some heat, the process can be made self-operating and self-perpetuating if a certain amount of heat is put in, perhaps at the beginning. This might mean burning some of the initial methane to heat up the generator to bring it up to efficiency.

Caustic Soda

Having set some sort of scene, I would like to discuss some typical chemical processes. As I have already mentioned there are a large number of simple labor-intensive chemical processes available, all capable of development into a state where they can be used in rural communities. Those which can be directly applied to any locality depend on the local resources and the local needs. It is possible here only to give a very random selection of processes to illustrate the sort of methods which might be chosen. They should not be taken as the most useful or the most economically valuable in all circumstances.

A promising intermediate chemical technology is the production of caustic soda, which is a good example of an inorganic chemical which is of great value in village industries. Caustic soda is necessary for the efficient production of soap and has other valuable applications as well. Of course, in modern Western industry all caustic soda is produced by electrolytic processes, using common salt (sodium chloride) as the raw material, which is very convenient, and generating caustic soda and chlorine—the chlorine is a valuable by-product. This you can't readily adapt for small-scale rural production, but there is some demand for small electrolytic plants in a few countries like Ghana, where cheap hydroelectric power is available. Ghana, in fact, was not at all interested in our chemical methods of making caustic soda; they wanted small-scale electrolytic caustic soda plants and we did in fact design some for them. The result, I might say, has been that they are wondering whether they should go back to the chemical process again, because it seemed a little difficult to get the communities who have these plants to understand electrolysis, whereas they can understand pouring chemicals into vessels and stirring them.

Although you can use this electrolytic process, it might be easier to go back to the earlier process, which is based on the reaction of slaked lime with sodium carbonate, a very old reaction. We simply have a slurry of slaked lime mixed with sodium carbonate. If heated together, you get a solution of sodium hydroxide and a precipitation of calcium carbonate or chalk—$Na_2CO_3 + Ca(OH)_2 = CaCO_3 + 2NaOH$. So this is a very simple reaction; you

heat a strong solution of sodium carbonate in an iron pot (which may be one of those oil drums I mentioned earlier) either over a wood fire, or by gas jets if local methane is produced together with injection of steam. It can be very useful to have a second oil drum which has a lid on it and a pipe going into the first vessel; and this is filled with water, which is boiled so that steam is directed into the vessel, which provides extra heating and what is more important, provides a method of stirring without having to put anything into a very highly caustic solution (figure 7.1).

When the reaction is complete a finely divided sodium carbonate exists in the vessel together with the solution of sodium hydroxide which can be filtered off. It can be filtered once again by pouring it into another oil drum which has had holes punched in the bottom and in which there is some sort of filter mat, ideally asbestos or something like it which is not decomposed by strong alkaline solutions. This filters the calcium carbonate off and leaves a solution of sodium hydroxide which can then be evaporated in iron pans. The iron pan can be an oil drum which has been cut in half, once again with some sort of burner, with a lid over the top to prevent air getting in and converting the sodium hydroxide back to sodium carbonate. You can either take this evaporation down almost to dryness, when sodium hydroxide will come out as crystals, or very often the concentrated solution is quite suitable for the next stage in the process.

Soda Ash

One of the problems, of course, is that we have only moved one stage back; we have produced sodium hydroxide from calcium hydroxide, which is easy to get, and sodium carbonate, which is not easy to get, because very little sodium car-

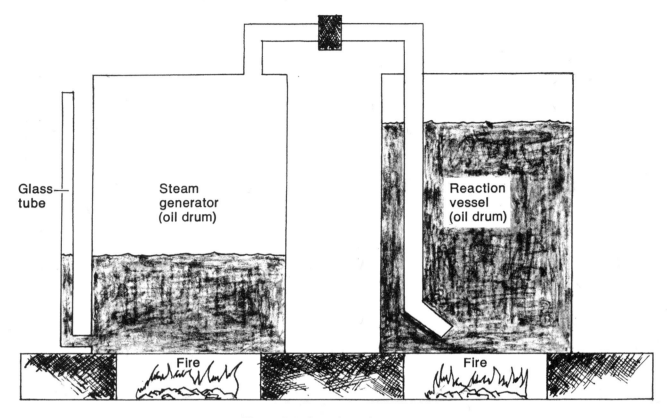

Figure 7.1. Caustic soda process.

bonate occurs naturally. This is often a difficulty, but in fact a number of plants do contain, or when carefully ashed can produce, soda ash which has a high concentration of sodium carbonate in it. A number of seaweeds represent important sources of sodium carbonate, as we will mention later. Other plants, cocoa pods, for instance, provide a very useful source of carbonate ash. Although in this case it is not sodium carbonate, it is potassium carbonate. When this is used for making soap, it produces potassium soaps which are liquid, but liquid soaps are very useful and can therefore be of considerable value.

Electrolysis

As I have mentioned, you can use a small electrolytic plant process to produce sodium hydroxide and it is being used in some places. A further use for these small electrolytic plants has recently become apparent. One of the great difficulties in using electrolytic plants in intermediate technology is that the conditions under which they have to be operated need to be very carefully controlled, otherwise the products of the electrolysis mix and you don't get sodium hydroxide and chlorine liberated; you get sodium hypochlorite. We know this in England as Milton, a very useful sterilizing agent and disinfectant.

Soap

Once you have caustic soda, it is very easy to make soap. You have only got to boil vegetable oils or fats with caustic soda for a sufficient period of time. If, as I've already mentioned, the caustic soda production and the soapmaking are carried out in conjunction, you do not need to produce solid caustic soda; the process-concentrated solution is perfectly satisfactory. Quite a lot of work has been done on this sort of process; a lot is going on in Ghana, a certain amount in Nigeria, and I was surprised to find in Pakistan how widespread is domestic soapmaking. You find families which are certainly outwardly quite wealthy but which by tradition have always made their own soap for washing clothes and still continue to do so. I got some very good recipes from Pakistani housewives on how to make soap; some of it is really very good, including certain perfumes and oils they put in it,

which produce fine soaps. I wonder why they ever buy any at all, except of course that some of it is loaded with caustic soda that takes your skin off.

Wood

Wood is a very important source of raw material. If you read the chemical literature of the nineteenth century and earlier, you find that wood was then a very important source of organic chemicals (figure 7.2). Wood gives us cellulose; we can get wood sugars, turpenes, perfume; we can get lignin from which we get chemicals and plastics; tannin extracts, wood gas; and volatile chemicals such as methanol, acetaldehyde, acetone, and

Table 1
Tannin Content of Some Plant Materials

Plant Materials	Percentage of Tannin
Chestnut wood	4-15
Hemlock bark	10-20
Tanbark oak	15-16
Chestnut oak	10-14
Black oak	8-12
Sumac leaves	25-32
Quebracho heartwood	20-30
Mangrove bark	15-42
Wattle (acacia bark)	15-50
Myrobalan nuts	30-40
Sicilian sumac leaves	25-30

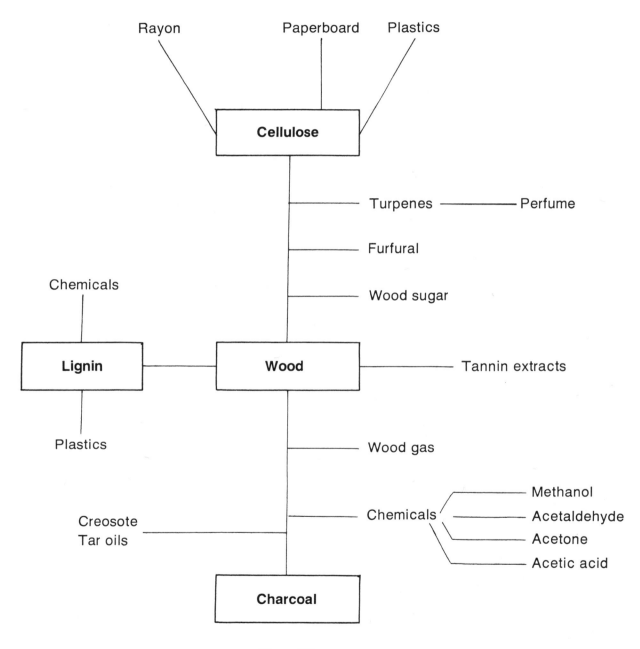

Figure 7.2.

acetic acid. We can also get charcoal. So wood as a source of raw material is highly important in intermediate technology. The sort of chemicals one can get in addition to those listed are methyl mercaptan, dimethylsulfoxide, vanolin, and oxalic and formic acid, and several phenols.

The value of distillation before the production of charcoal is therefore shown to be very important and if you look at a list of materials which you can obtain by distilling one ton of hardwood, it requires no further emphasis (table 2).

If you do not distil, but you treat wood chips or sawdust with dilute acid under pressure, you can get 40 to 45 percent of wood sugars. The requirement for pressure does tend to limit the value of this sort of technique in rural communities, but it certainly does not in small townships. In the Indian subcontinent the local craftsman is quite capable of making pressure vessels out of reasonably thick steel. These processes are well within the capability of local intermediate-technology industry of a slightly more sophisticated sort. Once

Table 2 Products Obtained by Dry-Distillation of 1 Ton of Hardwood Scrap	
Charcoal	270 kilograms
Gases:	150 cubic meters
Carbon dioxide (38%)	
Carbon monoxide (23%)	
Methane (17%)	
Nitrogen (16%)	
Methanol	15 liters
Ethyl acetate	70 liters
Ethyl formate	6 liters
Acetone	3 liters
Creosote oil	150 liters
Soluble tar	100 liters
Pitch	30 kilograms

you have wood sugars, you can produce a very valuable livestock feed and, having produced a wood sugar by treatment of wood under pressure, you can then ferment it to yield both alcohol and yeast.

Charcoal

I have mentioned production of charcoal. Charcoal is generally made by burning wood in covered pits which are so built that the entry of air is restricted and a charcoal burner produces an extremely good-quality charcoal. But the problem is that all the valuable materials, mentioned above, are lost. In places of high rainfall, or where there is a prolonged wet season, charcoal can only be made during a very restricted part of the year. This is a further argument for the use of distillation of closed vessels. I found that there is a great reluctance on the part of the population to use charcoal, once you have produced it for them, in spite of its greater efficiency than wood. This is because, if you make charcoal somebody has to pay for it, whereas, generally, there is an adequate supply of free wood in the forest, especially in Africa. So the demand for charcoal is a good deal smaller than might be expected. But the fact that all sugar production on any scale requires activated charcoal for cleaning up, which has to be imported, is a very good indication for trying to make activated charcoal by intermediate-technology processes in rural communities. This is not very easy, but it is one of the things that we are in fact working on. Rice husks appear to be a promising starting material.

Sugar

Sugar is a universal requirement, universally sought, not only in the underdeveloped countries. Small-scale methods of producing sugar exist worldwide and sugar is produced with varying degrees of efficiency in almost every country where sugarcane will grow, and it has been spread more widely with the spread of sugar beets in the countries where sugarcane will not grow. Sugar, therefore, is an important food. It is also a very useful source of many chemicals, and the bagasse which is left after the extraction of the sugar from the sugarcane is a very useful fuel and a source of building material. In fact, its value for making paper and building materials is such that, perhaps, we ought not to burn it, but ought to think of some other method of heating the sugar extraction process.

Whether the crop is cane or beet sugar, the sugar itself is the same. You purify it and get sucrose which is the cane sugar that we know and is also a food; other sugars such as glucose and fructose are produced in the refining process. In fact one of the big difficulties of intermediate technology techniques for producing sugar, is that it "inverts." In other words the molecule of sugar which is $C_{12}H_{22}O_{11}$, breaks up into two other smaller sugars—$C_6H_{12}O_6$, one is fructose and one is glucose—and these sugars prevent good crystallization of the sucrose and also have less food value. Also from sucrose you can produce polymers, drugs, and plastics. During the sugar-crystallization process you get molasses, which can be used for cattle feed, for making silage, or making organic chemicals, or for fermenting and then distilling as alcohol. As we know, fermenting molasses gives you rum, which is a delightful thought. We also have the hard parts of the cane's stalk giving us bagasse, which is used as a fuel, in fact in many cases as a fuel in the sugar process itself, and you can make paper and wallboard from it as well. So sugarcane is really a very important source of many materials. This is illustrated diagrammatically in figure 7.3.

Many villages throughout the world grow sugar and after harvesting the canes,

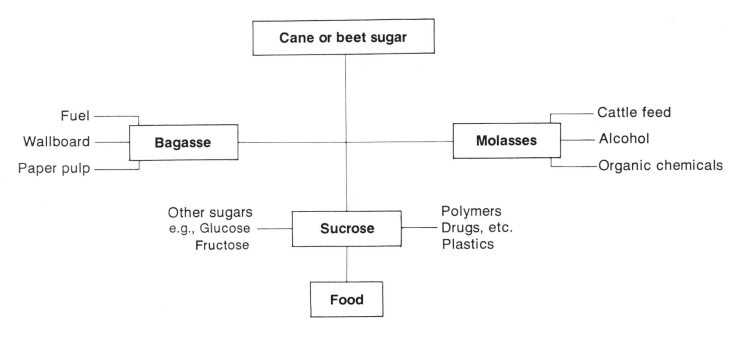

Figure 7.3. Some products of sugar.

the first thing that has to be done with them is to crush them to get the sugar out.

In a typical village sugar factory there will be a three-roller crusher. Its three rollers will be mounted on some sort of crude springs and driven by a pole which is either walked round by the women of the village or by an ox; they do not seem really to mind which, as long as the men don't do it. The juice pours down into a hole in the ground and is collected there in a pan. In places such as Karachi and other parts of

India this sugarcane juice is sometimes sold mixed with water as a drink in street stalls.

The first problem that needs tackling is that this is a terribly inefficient way of getting the juice out of sugarcane. Sugarcane contains about 14 percent sucrose and about 85 percent of the cane is liquid. So it is important to get a very large amount of this liquid out. The three-roller system, which is used because it is traditional and simple, especially the one

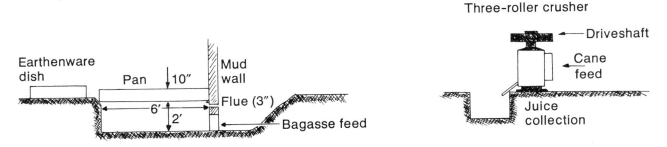

Figure 7.4. Village sugar factory.

111

turned round by oxen, loses a very large percentage of its sugar. One of the things on which we have been working is to produce much more satisfactory types of rollers. In fact you really have to go away from rollers altogether and to use some expression system with an Archimedean screw—a sort of mincing machine, so that the material is passed through a helical screw.

Once you have the sugarcane juice, it is a very sticky, thick gray juice which has to be purified. First it has to be clarified and it is desirable to reduce its acidity, because it is quite an acid juice. It is like many biological systems while they are actually a living system; they can tolerate chemical conditions which, the moment the material is extracted from the system, simply can't be tolerated at all. Sugar remains as sucrose in a highly acid condi-

tion inside the sugarcane, but as soon as you get the acid juice out of the cane it promptly starts this process, known as inversion, of breaking the molecule in half, losing a molecule of water and turning to glucose and fructose which is very difficult to crystallize or which stops the sucrose crystallizing resulting in a much less satisfactory sugar. So one has to clarify sugar. The usual way of doing this is to add lime—we are back to calcium carbonate again—and then for the best results sulfur dioxide which can be made on the spot by burning sulfur in air, is forced through the mixture.

Having obtained a juice which has been clarified, next you have to filter it. And filtration is another problem in intermediate technology. It is very easy in the large factory to filter efficiently, where one has vacuum systems and good filter pads;

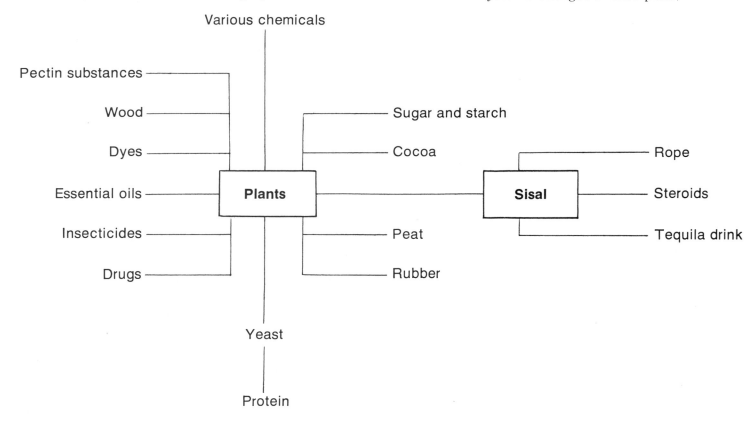

Figure 7.5. Plant products.

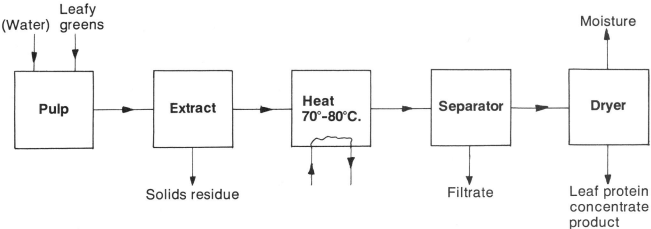

Figure 7.6. Leaf protein process.

it is much less easy in a village situation. So a lot of attention is having to be paid to methods of filtering without losing a lot of sugar.

Then the sugar has to be concentrated to a point where it crystallizes. This has to be done quickly because, as the temperature is raised, the tendency for the sugar to invert is very much larger; the higher the temperature, the faster the reaction goes. Now this problem is overcome in modern sugar making industries by using what is known as vacuum pans. The sugar is contained in a vessel, the space over which can be evacuated. Therefore, the boiling process is speeded up, because the vapor pressure is much lower and therefore the evaporation takes place in a fraction of the time it does in an open pan. However, it has not yet been possible to produce a vacuum pan which is suitable for use in rural villages and crystallization remains a major problem.

Traditional Processing

Rural villages produce a sugar known as gur, where the juice is simply extracted from the cane with one of the crude presses discussed earlier and then it is simply filtered, clarified a little, and evaporated. You end up with a product which is a dark brown, semicrystalline, rather sticky material. Unfortunately the sugar does not keep very well. It is difficult to store because of its sticky nature; it really is not very well crystallized and it does contain an enormous number of impurities. The other type of sugar available is kansari, which is treated with the lime sulphite processes and then filtered and evaporated under optimum conditions. A good, semi-white crystallized sugar is produced; this is the purest sugar produced in village communities.

I have found that this process is a very important part of industry in villages. Many of the indigeneous foods are based on sugar, which is a good source of energy but contains no protein, no minerals, and no vitamins. The sugar has no food value other than being the cheapest known source of calories, so it is useful as energy food. Another important use of sugar, apart from bagasse and its value as a fuel, is the wax within the cane. If you extract this wax, after removing the sugar, by treating with benzine in a solvent-extraction plant,

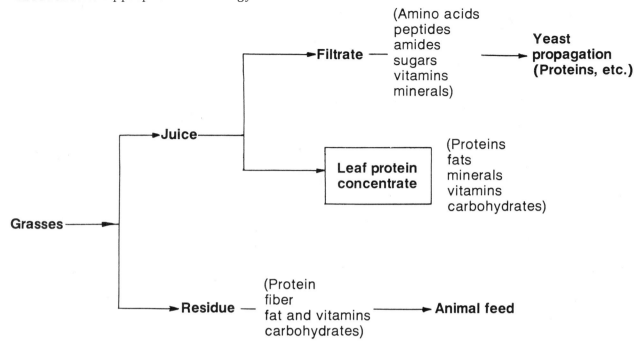

Figure 7.7. Leaf protein products.

it can be a very useful way for waterproofing or polishing. Sugarcane wax is something which we are trying to introduce into sugar-producing countries, because it is a raw material, which is, at the moment, simply getting burnt with the bagasse.

Plants as Raw Materials

Plants are a splendid source of raw materials (figure 7.5), providing yeast, protein, chemicals, sugar, and starch. Cocoa is a very valuable material obtained from a plant. Sisal plants will give rope and steroids and also tequilla, which people tend to forget. You can also get peat and rubber, insecticides, drugs, essential oils, dyes, wood, and pectins, which all can be produced from local plants. So plants are really extremely important sources of chemicals for intermediate technology processes.

Leaf Protein

One very important process, involving a local plant, or a very large range of local plants, is the production of protein from leaves. This is a simple process, which can be very valuable. Many green leaves, grasses, and herbage such as clover, can be treated to produce protein concentrates. All you need to do is to cut the material finely and then pulp it with some sort of press water (figure 7.6).

Having obtained a pulp, you can run off the liquid, which is fairly sticky, from the pulp. So you make an extraction of the liquid, which means that you are left with a residue of solid. The fibrous part of the plant can be used as cattle feed; it still has quite a lot of calorific value, although it has lost a lot of its protein. Then, if this extracted liquid is heated to 70 to 75°C., the protein coagulates into a white lumpy material, which can be separated from this by filtration. The product is a filtrate and a solid material, which can be dried to give a white powder, a very valuable source of protein. Although not all the essential amino acids are contained in leaf protein, it can be used as an additive to food in high-protein breads and in soups, stews,

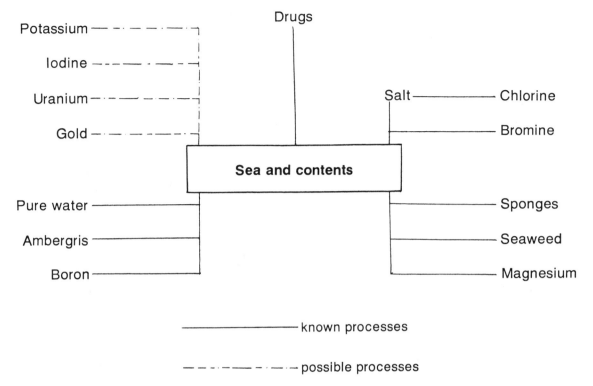

Figure 7.8. Products of the sea.

and curries to increase their protein value. In fact, up to 15 percent of the original weight of the grass is protein, so a very large amount of protein can be produced. Figure 7.7 shows the products of this process.

Other plants which are important are those, such as eucalyptus, which yield essential oils and those, like ephedra, which contain important pharmaceutical compounds (ephedrine). These can be extracted at village locations from locally

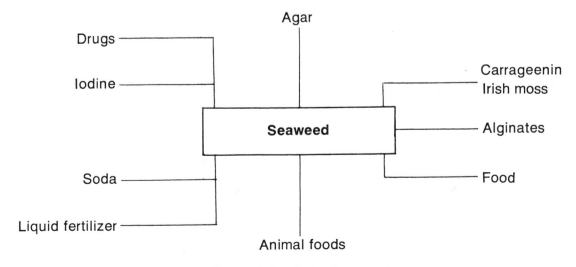

Figure 7.9. Products of seaweed.

115

grown, or gathered, material. The crude extract can then be sold to urban processing plants.

Marine Products

Seaweeds are also of great importance, as are other products of the sea. Figure 7.8 gives some indication of materials available in the sea, and figure 7.9 shows the range of products which may be obtained from seaweed. Precisely what may be obtained depends on the type of seaweed available, but the products yielded by drying, dry distillation, and water extraction of 100 tons of seaweed kelp are shown in figure 7.10. It will be noted that three tons of sodium carbonate are produced, which may then be used to make caustic soda.

Intermediate chemical technology holds exciting challenges for the chemical engineer who desires to explore the available alternatives. And with some fundamental knowledge of the chemical processes, people will be able to take greater charge of their own fate.

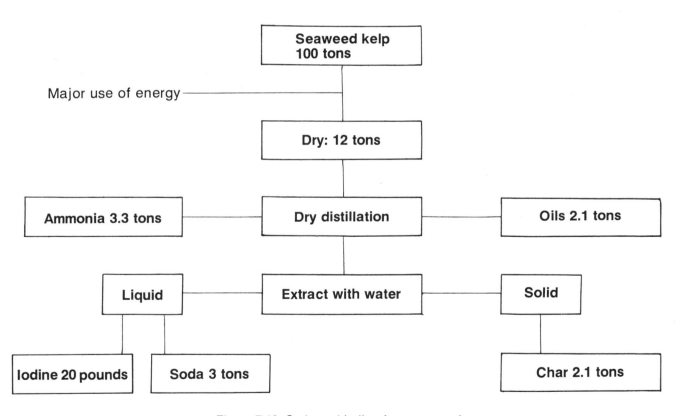

Figure 7.10. Soda and iodine from seaweed.

Chapter Eight

Education Systems: Appropriate Education and Technology for Development

by Charles R. Tett

There is no known human society in the world which does not have a system of transmitting its skills, wisdom, and way of life from one generation to another. As one African writer expresses it, "The claim that education was introduced to Africa by European colonial administrators and missionaries is not only terribly presumptuous but an indication of how contemptuous some Westerners have been of anything and everything non-European."

Dr. Anza Lema, another Third World educationalist writes of traditional education: "It might not have been as formally structured as the Western system was, but it was well organized and very successful in achieving its objectives." He states that there are three main educative agencies in African traditional communities.

1. The first was the family which was without a doubt the most fundamental.
2. The second was the informal lessons learned from contact with playgroups and the community as a whole.
3. The third more formal and organized part of the educative process was given during the years of adolescence in what Dr. Otto Raum calls "developmental rites." These usually center around the attainment of puberty and usually involved "circumcision rites."

After this intensive period of community education a person became legally free and qualified to inherit, to own, and to dispose of property. Young people were then given the right to marry as well as procreate and to raise a family. These rites drew public recognition to the new status of adulthood. This "schooling" period had educational significance in several ways.

1. It was a process of toughening the young initiates through various tests of physical endurance and ordeals of manhood or womanhood.
2. It united all the initiates together in a strong brotherhood by emotional and social bonds.
3. Social customs, religious beliefs, tribal law, and civic duties were taught by which they learned their obligations and duties to their community.
4. Skills were imparted by which the initiates were taught to build a home, care for the family, rear domestic animals, produce crops, fish, and hunt. The technical skills or making implements and weapons often came into this training period. The special skills of iron smelting, blacksmithing, medicine, and other specialized trades (arts) were usually passed on from father to son by life apprenticeship.

This traditional system varies from country to country and tribe to tribe according to the local environment, tribal structure, and customs, but it cannot be ignored as it was a valuable system.

Colonial Education

On this subject of colonial education let the Third World speak again. H. F. Makulu writes, "The expansion of colonial empires swept away the remaining strength of indigenous civilizations and planted in their place a new technological civilization . . . . The education provided by the Colonial Governments was not designed to prepare young people for the service of their country; instead, it was a desire to inculcate the values of Colonial society and to train individuals for the service of the Colonial State." This may have been done in some cases with the highest of motives for William Pitt in 1792 in a speech to the British House of Commons said, "Even Africa, though last of all the quarters of the globe, shall enjoy at length in the evening of her days, those blessings which have descended upon us so plentifully." Not all colonialists however, shared this high motivation, for those engaged in commerce were quick to see and exploit for financial gain the supply of cheap labor and raw materials which became available in the countries which they colonized.

Systems of education were set up by the various powers, the British, French, Portugese, German, and Dutch, all of whom had different approaches, not only to education, but also to politics and administration. It must be recognized that the colonial educators introduced the type of education with which they were most familiar and made little attempt to adapt it to the new environments in which it was taught. In all these colonial approaches to education the expatriate teacher saw himself as the source of knowledge and his pupils as empty vessels waiting to be filled. The role of the pupil was to be passive, docile, and never to challenge the authority of the teacher. This type of education is described in the Third World as "education for domestication." It regards the world as a closed system, which can only be known and understood in its totality by an educated elite whose responsibility it is to pass on these deposits of information intact. This knowledge is divided into departments called subjects, which are usually kept in isolation. Teaching is held to be a different function from learning, and the process of knowing is kept separate from working, and thinking is quite distinct from doing. It prevents any dialogue between man and the world. Such a position ensures that those in authority control knowledge for their own ends and thus perpetuates "the domestication" of the people under their wing.

Until recently the colonized people were never in a position to challenge this attitude. Indeed many of them have accepted the system because they saw new opportunities to acquire the secrets of European power and material benefits through the colonial educational system. From the beginning of the twentieth century this Western system of education was assumed to be the answer to economic and political progress by both the colonizers and the colonized. This type of education has benefitted the few who have been fortunate or wealthy enough to avail themselves of it but it is by no means the community education that they had inherited from their traditional system. Although quite often the two systems have continued to function together in parallel in the rural areas, in the fast growing urban centers they have been too difficult to continue.

The educational system, like all other imposed systems, has been challenged by the coming of political independence to the ex-colonial countries. Most countries, however, have been obliged to continue the same educational system because the already educated elite have vested interests in its continuance and are not prepared to consider any alternatives. Nonetheless, some countries have begun to challenge this system because of their ideological motivation and are putting a great deal of effort into research to find a system which will more suitably express their national "selfhood" rather than continue to operate a system which is foreign to their own culture. President Kenneth Kaunda of Zambia said, "The age of the old unimaginative approach to education is over and the time has come for all young countries to engage in serious reappraisals of their existing educational systems with a view to overhauling them entirely and gearing them to the needs of their countries."

Education and Culture

We have seen that when our own Western systems of education and training are applied to developing countries, all too often the local culture is ignored. Wittingly or unwittingly, a monoculture is produced. There is a tendency for modern life in all countries to be overly commercial and to overlook human values. In the cities of most countries it is hard to tell whether one is in Nairobi, Lagos, or London for the buildings are monotonously similar. Monoeducation and training is applied to the minds of people in most countries as monoarchitecture is applied to their cities.

Photo 8.1. Nambale Village Polytechnic (Kenya).

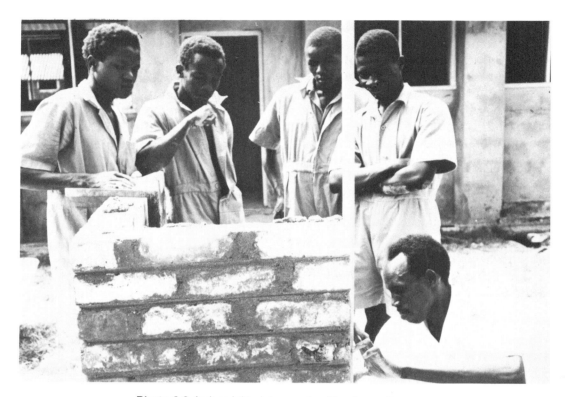

Photo 8.2. Industrial training center, Mombasa, Kenya.

Photo 8.3. Repair classes at Naro Moro Village Polytechnic, Kenya.

This is also true in the field of technology which does not take into account the local needs and environmental factors.

The "educated elite" usually have the same tendency to "keep up with the Joneses" as they are "hooked" by the same advertisements to buy Western consumer products and accept Western standards while complaining, not too loudly, that their culture is being destroyed. Influence from the West has succeeded all too well, and through the process of imposing Western education for many years, has produced people who have little relationship to their own cultural heritage. As a result they have been left in their "independent nations" as one African poet describes himself, "hanging in the middle way," belonging to two worlds and to none.

This pseudo-Western life-style, however, is only possible in the cities where there is money to support it for a small elite. It is not true of all developing countries, for even in Africa there are glorious exceptions (like Tanzania) where people are fighting hard to find real independence and to keep the African concept of living for the community and not for self-aggrandisement. There are no obvious signs yet that this battle is being won. In most developing countries the majority of the population is excluded from this process we call "development." Development is sometimes the name we give to the syndrome of our own imbalanced society by which the rich get richer at the expense of the poor, for everything is sacrificed to economic growth. True development must surely be the development of people; a process by which they are able to fulfill themselves as persons and make a contribution to their community. This cannot be measured in terms of g.n.p. or by a computer, for it is qualitative rather than quantitative.

Urban Unemployment

The plans for development in most developing countries have been based on raising the gross national product by the introduction of modern industry which is designed to save labor. The result is growing unemployment, especially in the cities. Unemployment denies the right of human fulfillment and its dehumanizing process only produces frustration, poverty, and misery.

Another factor which has exacerbated the unemployment problem is the increasing number of young people reaching employable age. This "population explosion" is taking place in the developing countries at three or four times the pace of the developed countries. Africa with 214 million people in 1965, will number 350 million by 1985. The male labor force will have increased by 50 percent by the same year, and will probably more than double by the end of the century.

Family planning is reluctantly accepted by people who have little family security. Large families are welcomed, as children soon provide extra working hands for the poor-yielding subsistence farms. They also are an insurance policy for the survival of the family in countries where medical services are in short supply and the death rate is consequently high. It is estimated that in India at least six children have to be born in a family to ensure the survival of a male heir. But even if family planning could be rigorously applied immediately, no effect in the labor force would be felt for many years, for the labor force is already determined for the next 15 years. The problem of unemployment is therefore a growing one. It is unlikely that the creation of new jobs will take place faster than the growth of the population unless more radical measures are taken to encourage a technology which is more appropriate to the needs of the developing countries.

Choice of Technology

It is apparent that little thought and planning went into the choice of technology that was originally adopted, or that any choice was ever offered. They have been sold a technology which matches the education system which has conditioned their thinking, rather than a technology which is appropriate to their needs.

This is a capital-intensive technology which has been designed for use in developed countries where labor is an expensive commodity and capital is readily available. Conversely, in developing countries labor is readily available and comparatively cheap, while capital is in short supply. It follows therefore, that if labor could be substituted for capital by the introduction of a more appropriate technology it would help to solve some problems. The appropriate technology that has been proposed is called "intermediate technology" for it lies between the traditional methods of the past, and the most advanced, expensive technologies of the present. It is a dynamic technology that develops with the development of the countries which apply it and leads to a fairer distribution of wealth and employment opportunities. Intermediate technology is more likely to keep the control of their economies in the hands of independent countries, encourage the maximum self-development, and reduce dependence upon the rich countries.

Experience gained in some developing countries shows that the application of modern or advanced technology in many cases actually creates more unemployment even when production is increased. In Kenya between 1954 and 1964 manufacturing output rose by 7.6 percent while em-

ployment increased by only 1.1 percent. This is typical of many other countries who have fallen for the same sales talk. These and many other factors contribute towards the increase in the numbers of unemployed in the cities and consequently result in increasing poverty and peri-urban squalor.

Rural Unemployment

The prospect for the majority of the 90 percent in rural areas is also one of unemployment, and, as there is a growing tendency to mechanize the large farms, the number of job opportunities in the agricultural sector is diminishing rather than increasing. There are, of course, a few places in the developing countries where mechanization produces more employment, but these are pitifully few.

The majority of the people depend on subsistence farming which helps them to maintain a mere existence. In many areas the small plots of land and poor marketing facilities do not produce sufficient cash crops to improve their standards of living. Usually their low cash income is barely sufficient for the payment of the obligatory government taxes. It is no wonder that young people flock to the cities even if there is only a remote chance of getting some cash-producing employment. The primary school education which they may have received in the village school has done little to prepare them for life in the city. A few of them have their expectations raised by being able to continue to secondary education, but even this does not necessarily lead on to employment.

Is Education the Answer?

Formal education has all too often been regarded by many young politicians in developing countries as the panacea for their countries' problems, and young governments have spent disproportionate amounts of their budgets on the type of education that is relevant to the needs of the more developed countries where conditions are entirely different. It may be that because of political independence it was necessary to train as quickly as possible those who would replace expatriates in the public service and staff management positions, but this situation is over for many countries. The education system needs therefore to be reviewed to meet present needs or there will be a surplus of trained and educated young people whose aspirations will be frustrated, and which could have serious social and political consequences. In the Kenya development plan of 1970/74 it is stated that of the 138,000 secondary school leavers, 70,000 of whom will have completed four years' secondary education, perhaps less than 50 percent will obtain wage employment. The plan states, "The only group who will have little difficulty in obtaining wage employ-

ment will be those who have received technical or vocational training." But increased opportunities for technical training will not answer all the problems, although in urban areas it could provide more job opportunities. I have not singled out Kenya for any reason other than it is typical of many Third World countries who are trying to adapt the Western system of education to their needs. Similar difficulties are being experienced in most developing countries and a great deal of rethinking is taking place with a view to reappraising the educational systems.

The education plans of the Indian government should provide a sufficient warning to others. Their fourth five-year plan states with some pride, "During the course of the third plan there was a considerable expansion in facilities for engineering education at both diploma and degree levels." But in the year that this was published (1969) 45,700 engineers with degrees and diplomas were put onto the

Photo 8.4. A diet talk at a local hospital in Nigeria.

Photo 8.5. Masai Chief shows his cabbage, Rural Training Center, Kenya.

Photo 8.6. Learning rabbit rearing, Kenya Village Polytechnic.

labor market already overflowing with 20,000 unemployed qualified engineers. Yet sanction has now been given to train 72,000 more every year.

Education by itself cannot solve the problems of unemployment unless it is appropriately planned and applied to a technology appropriate to the needs of the country. We have to accept however, that modern cities exist in Africa, as do modern industries, and that there is still a need for formal education and training to meet their special requirements. The question is, what is "appropriate" education and training to meet these needs?

Formal Education and Training

The present pattern of formal education fits the formal system of modern commerce and industry for which it was designed. But this sector of the developing world represents only a very small portion of any country geographically, and likewise employs only a small percentage of the population at present. Both the academic and technical education systems adopted by these countries therefore, apply only to a small number of people and produce an intellectual and technical "elite." The majority of the population remains unhelped by the wealth created by the modern formal sector of industry and commerce. There is a good deal said by some economists about the benefits of the modern sectors being distributed to the poor rural areas, but there is little evidence of this. Academic education is overproducing arts graduates in most of these countries who expect a city job, and a high standard of living, but more of these young people are finding their expectations

Photo 8.7. Masai craft work on sale to tourists in Kenya.

frustrated. Insufficient emphasis has been placed upon expanding technical education to keep pace with the expansion of modern industry and its increasing complexity. If developing countries are to maintain the modern capital-intensive industries they have already started, formal technical education to maintain their industries must of necessity be supported by an increasing budget to enable it to keep pace. If, and when, there is a change of policy concerning the adoption of a more appropriate technology then the system of technical education will need to be modified accordingly.

The formal sector of education at present has little relevance to the needs of the rural areas except in the agriculture and veterinary services. There is still a need for this type of specialist training to expand, as the economies of most developing countries depend upon the products of the rural areas for food and raw materials for their industries and for export. At present the cost of educating and training a minority absorbs most of the national education budgets and little is left to experiment with more relevant types of education especially for the rural areas.

Appropriate education and training for rural living could increase the production of more food and raw materials and this alone should provide an incentive to any government. The increased yield of products and cash income would raise the standards of rural life and, though such progress may be slow, its contribution to the development of the nation would be invaluable in the long term.

The Informal Sector of Industry

In the "downtown" areas and in the marketplaces of rural areas there is a great number of informal small industries operated by entrepreneurs, some of whom have had no formal training, and some who after learning an artisan skill have begun their own businesses. These entrepreneurs are more numerous than one would think. In Nairobi, Kenya, it has been estimated that more than 10,000 workers operate in the informal sector and sell their products or services to support themselves and their families. Training for self-employed technically skilled men and women is almost nonexistent in many developing countries, yet the opportunities are great. It has to be recognized that entrepreneurs, like artists, are usually born not made, but both need training to maximize their potential. This is as true in Birmingham as in Bombay or Bungoma, for British industry still depends on small entrepreneurs to a surprising degree. Peter Marris in his research published in *African Businessmen* highlights the need for on-the-job training services to make these men more efficient and effective.

The type of industry most suitable for the informal sector usually follows the principles of "Intermediate Technology" which can be defined as:

1. Substituting labor for capital, thus providing more employment
2. Simple rather than sophisticated using the minimum amount of machinery and tooling of an unautomated pattern
3. Being inexpensive it makes minimal demands upon national imports and uses locally available materials as far as possible
4. Training for these industries is comparatively inexpensive as maximum use is made of apprenticeship and on-the-job training for imparting skills.

This type of technology applied to the informal sector in developing countries should receive every encouragement for it is from these roots that the truly independent and indigenous African industry can grow. This kind of industry using similar technology was the foundation of modern industry in the developed countries and is still viable today in many parts of the world, especially in the developing countries, which are at the same formative stage in their national history.

The Intermediate Technology Development Group has been able to demonstrate this approach to technology in Zaria, at the request of the North Central State of Nigeria with aid received from the British Overseas Development Administration.

A workshop has been established in which prototypes of hospital, school, agricultural, and other equipment have been designed and tested. A training scheme for unemployed youths has been established for the production of this equipment, using simple procedures and labor-intensive methods. There is now an increasing demand for the products for local consumption, also for the services of the trainees when they have completed their training. Some of the ex-trainees have been able to start their own small enterprises. After seeing the benefits of the scheme, other Nigerian states have asked for similar help to establish their own intermediate technology workshops and training schemes.

Nonformal Education and Training

While it is necessary to give help to the informal sector of industry through training programs, there is also a strong need for nonformal education and training programs in the urban and rural areas.

It is not possible for any African country at present to give formal education at primary school level to all of its children, because they cannot afford it. Not only are the costs of formal school rising faster than national budgets, but the investment in schools has not paid off in providing employment for all the pupils who have completed their education.

"Nonformal" education and training programs are therefore needed to help the increasing number of unemployed to become productive. These can be divided into three main groups:

1. Those who have not had the opportunity of formal education.
2. School-leavers who have need of additional training to become employable.
3. Employees who need upgrading to enhance their productivity.

1. The Unschooled

It cannot be accepted that those who are illiterate are necessarily unintelligent, but simply that they lack the opportunity or wealth to be educated. For those who cannot be formally educated "learning by doing" is more applicable to rural living and meets the needs of a great proportion of "nonschooled" young people and adults.

This method can be linked to advantage with functional literacy schemes, or adult education courses where these are available. For the purpose of upgrading agriculture in the rural areas from subsistence level to better standards of food production and cash crop introduction, learning by doing is quite adequate and effective. Many other rural craft skills can be learned to a high degree of competency without the use of written materials or expensive modern teaching facilities. There are many examples of nonformal apprenticeship training from many parts of the world where a successful farmer or master craftsman communicates his expertise or skill. A trainee can be taught to accomplish a skilled task by personal tuition and, through experience gained in subsequent practice, can become competent and efficient and reach a high degree of skill and success. This type of training for employment needs active encouragement and where it is absent, every attempt should be made to introduce it.

A valuable example of this "learning-by-doing" method is provided in an unusual agricultural scheme in Nigeria which for many years has provided apprenticeship training for young farmers. Starting in a subsistence farming area the initiators of this scheme realized that hand hoes alone could not produce an agricultural surplus which could substantially increase the villagers' income. The use of tractors was out of the question for the capital required would cost more than the total income of the average agricultural worker for 100 years. Therefore, the people decided that the scheme would be based on the use of an intermediate technology using animal-drawn equipment, improved seed, and better agricultural methods. Starting with a small group of farmers, training was given by learning "on the job." Instructors introduced new methods of training in the use of animal-drawn equipment. When the trainees had reached a sufficient level of competence, they were given a loan of $200 to buy their equipment, bullocks, seed, and fertilizers with which to start their own new farms. In return they were asked to make two contributions. The first was a promise to help the surrounding farmers by demonstrating to them what they had learned, and secondly to train two young apprentices for two years. When these apprentices proved their worth, they were offered similar loans after making similar promises. The result of this "each one, teach two" has proved itself over 10 years, and 80 percent of the farmers so trained are still farming. When visiting I was told that all the farmers had repaid their initial loans except for a very few who had suffered some calamity.

2. The Unemployed School-Leaver

One of the most urgent needs in developing countries is for training the unemployed school-leaver. Formal education has usually been acquired by the sacrifice of poor parents and has raised the pupils' aspirations towards employment in the modern sector with the prospect of a secure future. There is considerable frustration when the majority are eliminated by competitive examinations, or by lack of fees to complete their education to university level. Pupils experience even more frustration on completion of primary or secondary education when they find out their education does not necessarily ensure for them a place in the modern sector of commerce and industry. If they are to be employed, further training is necessary after completing their schooling.

Many types of nonformal training programs, most of an experimental nature, have been established in Africa with varying degrees of success. Many have been set up by voluntary agencies and have depended upon large inputs of external aid to initiate them and in many cases to continue them. In the report "Non-formal Education in African Development" James Sheffield and Victor Diejomaoh describe a great number of these schemes in some detail, and show a wide variety of suitable approaches to the problem. One of the facts noted in their report is that these schemes are less expensive per trainee than formal education, but that in spite of this, there is a reluctance by most African governments to support them financially to any great extent, even though a high percentage of their national budgets is spent on formal education.

Sometimes it has been left to government departments other than the education department to encourage nonformal education. In Kenya the Ministry of Co-operatives and Social Services is aiding the development of the "Village Polytechnic" scheme which was initiated by the National Christian Council of Kenya, following their survey of school-leaver problems published in the report "After School What?" in 1966. In this report urban and rural youth problems were examined and it was realized that rural youth needed special attention and training in skills other than only agriculture. The term "Village Polytechnic" was coined by the report committee and the seed of a new type of training for village youth was sown. This seed has taken root and produced a number of different training ideas, which have grown as they have been fed to village youth leaders. As a result two types of Village Polytechnics are flourishing which can be classified as institutional and extension oriented.

The institutional type has drawn together groups of young people of both sexes and simple training has been given in various crafts by employing artisans or other qualified instructors. Short courses in poultry and beekeeping have made young people realize that there is an extra income to be gained from using to the full their local natural resources.

The other type of Village Polytechnic has become extension oriented and relies upon experimenting with new cash crops and the use of animal-drawn agricultural equipment. Some V.P.'s have combined both the institution and extentional approaches. The results have been encouraging. One V.P. claims to have produced in one year:

89 beekeepers with an average extra income of 1,200 Kenya shillings
79 new poultry keepers
7 new fish ponds, and to have been responsible for sinking 80 wells.

On a recent visit to one of these Village Polytechnics I was encouraged to see that the apathy that is often apparent in villages had been replaced by enthusiasm and a willingness to absorb these new ideas.

In the urban areas there are a growing number of experimental training schemes for unemployed urban youth. These programs usually offer training in artisan skills for work in industry or commercial training for office work. The demand for such training always seems to exceed the possibilities of supply.

3. Employee Upgrading Training

This type of training assumes that the worker is already employed and therefore, it is to the employer's advantage that his performance is increased, as this should lead to a higher productivity. There is an understandable reluctance on the part of the employer to pay for such training, as the more highly skilled an employee becomes, the more employment opportunities are open to him with rival employers, who are prepared to offer higher wages. In a developing country where technical skills are in short supply the risk of losing a technically skilled worker is great. The "levy" system adopted by British industry and by a few developing countries ensures that the financial burden of upgrading training is borne by all employers of skilled personnel. This scheme offers an incentive to all employers to upgrade the standard of their skilled workers which is both advantageous to them and does not place a further burden upon government funds. Upgrading courses can be organized by employers collectively or by attendance at special training centers funded from the money received from industrial levies.

There are of course many other nonformal education and training schemes but these examples have been chosen because they would seem to be the areas where immediate help is urgently required to stem the continual rise of unemployment. But although these will make a

Photo 8.8. Learning to build ovens from paraffin tins in Uganda.

Photo 8.9. Removing the baked bread from the ovens.

130

Photo 8.10. Water storage tanks above a borehole in Kenya.

Photo 8.11. A church training project in Nigeria.

valuable contribution to developing countries they are not a substitute for formal education, but supplementary to it. They should however, influence formal education to adopt a curriculum more relevant to the needs of national development, and remove some of the prejudice which seems to exist against manual work, however skilled, and however essential it is to the interests of national development.

If the developing countries are to find a system of education which will satisfy their social and economic aspirations, they will need to expand considerably their educational experiments and research. Each country must find its own way to fulfillment and this path will be neither easy nor short for there are vested interests at stake both internationally and within

their own nations. The present "elite," where established, will not find it easy to put national interests before personal ones. International governments through their trade and aid programs will bring considerable pressure to bear when their profits are threatened.

Such attitudes of self-interest can only be overcome by some other motivation. For, however appropriate the choice of technology, however good the educational system adopted, success depends upon the willingness of people and their leaders to accept change, to adopt new concepts, and to face the cost. This will not be easy but it will help to establish the national "selfhood" which they all seek and give them a confidence to plan their own future development.

Dr. Uga Onwuka writes, "The African needs a dynamic educational system that will foster self-realization, self-confidence, and promote cooperation and productivity. It should be a system in which all educational institutions positively serve national interests and thus offer the African the identity that he now lacks."

Summary

It is evident that the choice of an education system to suit all the "learning" needs of people living in the rural and urban areas of the Third World is not an easy one and some of the following factors must be taken into account.

1. Education in the widest sense must lead to "liberation and humanization" of all the people in Third World countries.

2. Informal education in which all that is best in their past tradition and culture should be retained and built upon.

3. Policies must be adopted that promote a growth of employment that will keep pace with the growing population. In this respect the choice of technology is a crucial issue if new employment opportunities are to be generated.

4. Education for employment will contain both nonformal and formal systems working in parallel and providing a comprehensive learning system, in which the elements of "learning by doing" and "symbolic verbal encoding" should be incorporated.

5. There will be a tension between the widely differing needs of the rural and the urban areas, but both areas must have learning systems that will suit local needs and enable all the people to make a satisfactory contribution to their nation and receive an equitable return.

Photo 8.12. Cinva Ram block-making machine.

Chapter Nine

Appropriate Production Systems: A Conceptual Framework

by Ben van Bronckhorst

When confronted with the complex phrase "socially appropriate technology," we might first ask ourselves what it means. Very simply the phrase refers to a technology that fits into the social pattern, a technology that, in contrast to the one we now know, has no damaging effects on men.

Perhaps this "perception" appeals to us because lately there has been considerable criticism of technology in the West. We are daily confronted with warnings about environmental pollution, we hear of the danger that threatens us from the large chemical industries which are usually located near living areas. We hear of the exhaustion of resources, the oil crisis, the food crisis, and so on.

I won't go further into this, because these subjects are already so strongly aired by the Western media. It is, however, important to mention this association because appropriate technology is one of the answers that has come up lately, one of the solutions which is being developed for these very problems. Mind you, in my opinion, appropriate technology is not the only solution; on the contrary, I am convinced that we could never tackle such a complex problem as the development of the world with one single solution. I attach such importance to appropriate technology because I have asked myself what we as technicians and designers of apparatus—or systems as they are called nowadays—can contribute to alleviating the problems mankind struggles with. Must we change our approach? And if the answer to these questions is positive, what must we do, what must we learn, which approach should we then follow?

I think that the phrase "socially appropriate technology" itself indicates the direction we should take. That is to say, socially appropriate technology must not be seen as a kind of remedy against all evils; far from it, but it may give us a method which enables us to solve at least some of the problems. This method, nevertheless, leaves us with certain obligations. The realization of a socially appropriate technology depends in the first place on our ability to make technical designs and to make them such that they mean a real improvement in a social system.

Technology

From a somewhat vague notion of the meaning of the word "social" and appreciating a technology that is good for mankind, we have arrived at a method—a design method—that can generate solutions for part of the problems mankind struggles with. It's time now to get a better understanding of what is meant by those three words that together form the concept. Actually, there are two main elements, society and technology, and the relation between the two must be appropriate. We'll start with asking the question: What is technology? When we consult the dictionary we find that technology is: "The science of industrial techniques, the study of the development of industry."

This definition is comparatively old and nowadays, under the American influence, we tend to treat the concept of technology differently. It now partly serves to indicate a certain field of our applied knowledge such as textile technology or medical technology. Generally technology is used with the meaning of: "The application of scientific knowledge for the solution of practical problems." Hence the somewhat casual use of this concept nowadays. When we speak, for example, of transport technology we mean to indicate that practical problems in the field of transport are solved with the aid of our scientific knowledge.

But this naturally means that we are gradually surrounded by the results of this process; all the solutions that have in the course of time been found for a whole range of practical problems continue to exist for a shorter or longer period. Therefore we have to take another step in order to make the concept of technology completely clear: "Technology is the sum of the solutions that have in the course of time been established and that have been integrated into the existence of men."

It is this last meaning which is by far the most important to most people on earth who have had hardly any part in the establishment of the solutions, nor will they have in future. Technology, in the sense of finding solutions with the aid of scientific knowledge is, certainly in the West, a matter for trained specialists, those people who have completed a scientific technical training, such as engineers.

Social Aspects

I believe it is important to be fully aware of these two different meanings of technology, for then it is clear why the term socially appropriate technology has been devised. For the engineer it means that he has to find solutions for practical problems, but for most people it means that they will have to live with his solution, and all the earlier solutions. It is therefore not surprising that both groups of people judge the solution differently. The main reason why there is an interest in, and demand for, socially appropriate technology is, in my opinion, the fact that the posing of the problem, the practical problem the engineer is going to exercise his scientific knowledge on, is too narrowly defined.

Certainly, a lot of attention is paid to technical aspects and economic implications, but we have many examples that show us that the attention paid to the social aspects has been very slight. Many experiments are done before a solution is released, but as engineers we can't predict very accurately what effects the complete workpiece will have on society, when many people have to work with it. We are frequently unable to predict what society will do with the solution, whether to judge this as favorable or unfavorable, whether the end result will be useful or damaging. Engineers do realize very well that the possibility exists that a technically and economically sound solution may be socially inadequate. But they haven't sufficient means to test the designs according to social criteria; in only a few cases is the knowledge of the social sciences adequate to pronounce upon this.

But it is a long-felt wish in universities of technology to subject designs to the test of social effects. Socially appropriate technology, if we may now use this term, has been at the top of the list of necessary developments for a long time.

I mentioned before that the reason why unexpected side effects occur in the implementation of technical solutions lies with the problem being too narrowly defined. Undesirable phenomena occur, sometimes at the most inconvenient moments and far removed from the object of solution, because our technology is really a system—a whole of mutually related parts. I have called this system a technological network, indicating that it is a special kind of system; transactions take place between the elements. Economists have already discovered this network; they speak of columns and branches in industries. And more recently, they have input-output analysis where money transactions between all sectors of a country's economy are represented in a matrix. Input-output analysis makes it possible to calculate what affects on the whole economy (e.g., in the development of the national revenue)

can be expected from measures that are executed in certain sectors.

We can extend this analysis beyond financial transactions to follow the movements of capital goods, personnel, materials, energy, and information. But such an analysis can't actually be carried out, not because we lack the processing apparatus, but because we lack the necessary data. However, we can give an impression of the technological network as shown in figure 9.1.

From figure 9.1 one can deduce that the balance in the whole is disturbed when

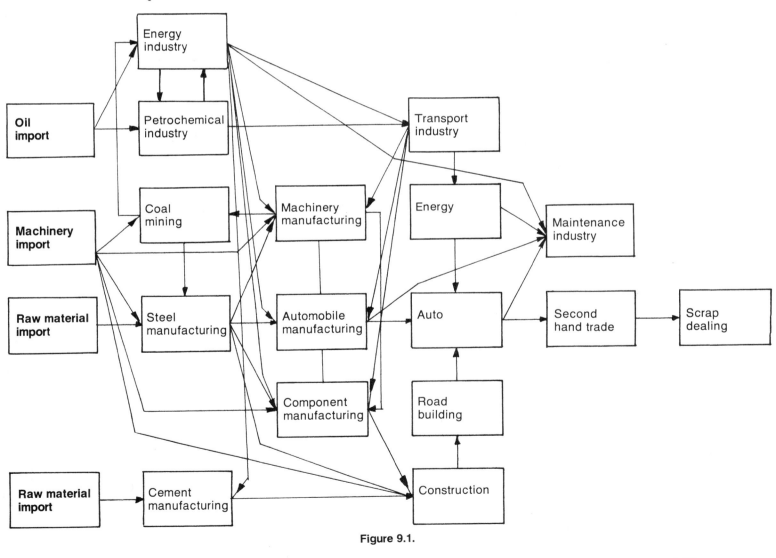

Figure 9.1.

measures are taken locally to increase the intensity of a particular stream. After a shorter or longer period this has consequences elsewhere in the network, causing a change in the relative scarcity of transaction objects, or causing a change in the loading on the sectors, which influences the efficiency. In most cases these disturbances will propagate themselves throughout the whole network, and as ill luck will have it, it is these disturbances that give rise to the definition of practical problems, which in turn leads to scientific design activities. It may therefore be expected that these disturbances won't solve anything, because the changes that are induced elsewhere again lead to new disturbances. Because solutions usually only result from partial considerations, side effects will continue to occur. In some cases a number of these effects will be negative at the same time and we then speak of a crisis. When such a crisis occurs the political conditions are often of such a nature that it is possible to have coordinated repair actions; the balance of the network is restored by drastic measures that affect the whole.

I have gone into this in some detail because it is necessary to understand that technology has at least two aspects: the aspect of design that the engineer sees, and the aspect of usage that most people are concerned with. And as stated before, both aspects are subjected to different judgement criteria. Technology is, moreover, the complex network of sectors of the economy that mutually supply each other and that form the basis of human existence.

People find their jobs in the various sectors of this network, and the continued existence of the network assures them of the availability of the necessities of life. A socially appropriate technology has three meanings: design, usage, and the basis of existence. The emphasis lies on the condition that no conflict may arise between

technology and social values. Technology must be subservient to life.

Technology and the Third World

In the foregoing section we haven't questioned that there might be differences between different countries, as to their respective places in the technological network. We can think of this network as spread out over the whole world. The accents differ, however, for the countries of the Third World mostly act as suppliers of raw materials, food, materials, and energy, and as such they are located at the beginning of the network. These countries, moreover, contribute only a small part of the material output; buying finished goods produced in the West. The larger part of the network is situated in the West; the developing countries occupy a peripheral position.

Naturally, the developing countries also have their own networks, which are traditional and directed at providing for their own populations. Although a similar analysis could be made of these networks, it would be even more difficult, not only because it would be hard to realize the processing, but especially because the data can't be acquired. Acquiring information is very difficult in practically all developing countries; only occasionally are there reliable data available.

In order to get an insight of the networks such as do exist in the developing countries, we will have to make-do with qualitative descriptions. And this is the first obstruction to doing research; one is used to quantitative data and now there is this sudden switch to qualitative data. Yet I'm afraid that we won't be able to escape this dilemma for some time, and we shall have to get used to simultaneously handling statistical information on Western

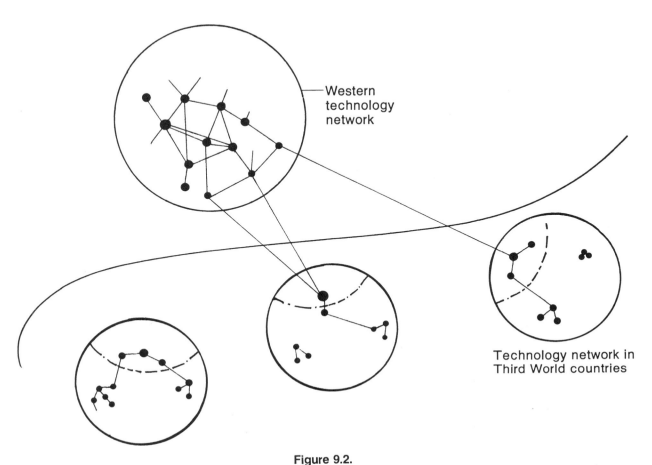

Figure 9.2.

networks and verbal descriptions of the situation in developing countries. Such descriptions already exist, but it takes practice to read them well and especially to be able to discern the data that are hidden in the text.

But I have given an impression of this point in figure 9.2.

There are two kinds of networks—one Western worldwide network, and a great number of local networks. The Western network has a comparatively uniform logic, and is transparent, while the networks in the developing countries are hard to see through, are complex, and appear to have been composed according to a different logic. The Western network is comparatively powerful—for various reasons—

and is able to exploit parts of the networks in the developing countries. Such an exploitation invariably means the destruction of the local network, for its balance is completely disturbed and these countries lack the institutions that could restore the balance again. To put it even more strongly, in many developing countries the existing institutions have such a foreign outlook that they tend to further the breakdown of the local networks, or to support the exploitation by taking various measures. Hence it is possible that some industries settle in developing countries because the presence of certain conditions is favorable. These industries are eagerly taken up by the developing countries for, from their standpoint, they also stand to gain by them. But in most cases the industries will remain on the periphery because in the developing countries the infrastructure that is necessary to integrate the industry is nonexistent, and cannot be easily developed. Therefore there often arises a situation of the type which can best be illustrated with data acquired from research in Indonesia, shown in figure 9.3.

As a consequence we find three barely related types of activity in developing countries.

1. Primary industries that export the raw materials to the overseas network, next to commercial enterprises importing expensive Western products to the local market.
2. Enterprises in the secondary sector, largely founded as joint ventures and to be regarded as transferred parts of the Western network.

3. Local supply systems, often very primitive and in many cases already affected by the influence of articles from affluent countries.

Each of these three types of activity have their own rules, so that with some justification one could speak of three separate economies. For the time being, however, I would prefer to stick to a twofold division, laying the dividing line between the part that is orientated towards the West and the part that is inherent to the country concerned. The first sector is rich, a lot of money is involved, and good jobs can be found there, and in both organization and working method it shows a great similarity to the activities we know in the industrialized world. The second is poor, with traditional working methods. And the larger part of the population is dependent on it. Because of the existence of this

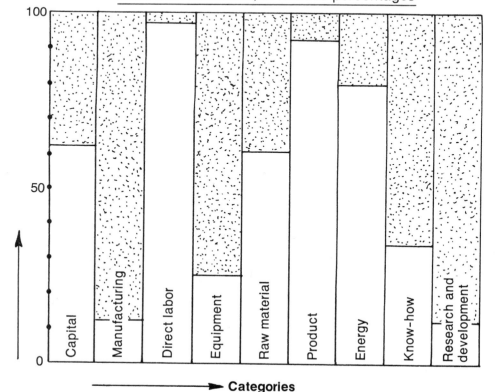

Data from Indonesia, in relative percentages

Figure 9.3.

separate "poor economy" the people in the Third World have been able to remain relatively free from disturbances and crises in the Western network.

This poor economy is an important element in the explanation of the paradox that, although the growth of the Western network completely lags behind population increase, still no major disasters have occurred. No matter how poor, this type of activity is still the best guarantee for the continued existence of the larger part of the population of the Third World.

The Poor Economy

The existence of this poor economy was noted long ago by a Dutchman, Professor J. J. Boeke, who first drew attention to the existence of two separate economies in what we now call the Third World. He was, however, pessimistic in his conclusions: the dualistic economy, the existence of a rich superstructure and a poor substructure, would not disappear so easily. The principles of the economic dealings of Western economies couldn't be applied to the "Eastern economy." The poor economy had its own logic, which seemed to resist every improvement. Since World War II development economists have devoted themselves to the economic growth of poor countries, and they understand very well that the difference in conditions compared with Western countries demands a different approach. On the one hand, they saw the rich countries with a large surplus of capital, relative scarcity of labor, and a rising buying power; on the other hand, there were poor countries with a shortage of financial means and a large, very poor, population. The technologies in each group of countries would have to differ according to the relative scarcity of means.

In the developing countries low-cost, labor-intensive solutions would have to be

looked for and the point was to choose the right technology. It was assumed that the wish to realize a large material output held good for the whole world, so that the same product would have to be manufactured under different circumstances. In the developing countries this would have to be done labor intensively. Because of the rising buying power in the developing countries, comparable goods would have to be manufactured in large quantities, and so production would have to be enlarged. This would need to be done partly in the developing countries themselves, and the designs of the production systems would have to be such that the special conditions prevailing in those countries were taken into account: an appropriate technology needed to be developed.

It was undoubtedly a reasonable aspiration to hope to close the gap, but that has not happened; everything changed, and faster than the development in the poor countries could progress.

- In the first place there was a strong development in the automation of the Western network, which led to a strong increase in productivity. And it was all the more strong because the infrastructure, too, underwent great improvements: the supply of materials, the provision of transport, energy supply, education, regulation of the financial market, etc. This made it possible to keep large parts of the network at home; technological development provided solutions to all practical problems that occurred.

- Secondly, the buying power in the West increased beyond all expectation, and the investments in the developing countries were used in the first instance to supply the foreign markets. The need for raw materials strongly increased too, and this led to an increase in activity in the rich sectors of developing countries.

• Finally the product itself underwent a stormy development, many new manufacturing possibilities were opened up with the development of the modern, largely automized technology. The products were imported by the developing countries, the mechanically manufactured product partly replaced the locally made one, but also new products appeared on the market—electronic apparatus, household articles, and especially motorized transport.

These products couldn't be manufactured locally and no efforts were made to look for labor-intensive solutions. It is indeed doubtful whether, given the previous arguments, such solutions could have been found.

So it was only possible to a very small extent to compete "labor intensively" with the products from the West. A further important obstruction in establishing alternatives was the entrepreneurs themselves, who, when put to the choice of a tested, capital-intensive technology and a still-to-be-designed, labor-intensive method, clearly preferred the former solution. Labor-intensive methods are therefore only found in a few cases in the rich sectors of the developing countries.

Production Functions

I will further elucidate this with an example, but first I want to give a little more microeconomic theory. It is assumed that capital and labor together determine the output. The output is a function of the inputs or production factors of capital and labor: $0 = f(C,L)$—at least for a certain state of technique.

Technical development means that a different relation arises between capital and labor on the one hand, and output on the other hand: $0 = g(C,L)$ where $g \neq f$. In the West we limit the concept of technical development to those cases in which the output is increased with the same or a smaller input of capital and labor. The relation between output and input is called the "production function." The production functions f and g specify a large number of combinations of capital and labor that produce an equal output.

It is usual to represent production functions as continuous, or to put it differently, all combinations which are specified by production functions are equally realizable. Which one will be chosen then only depends on the price of the production factors. One endeavors to realize a maximum output with a minimum of costs, the cost per unit is then minimal. This produces a further restriction in what we in the West consider technical progress; for not only is it a matter of increasing the output with an equal or a lesser quantity of input, but also the price of the input has to remain the same or be reduced. This means that the direction of technical development is determined by the price development of both production factors. Of various suggested innovations only those are realized that fulfill these conditions: the output must remain at least equal, but preferably be increased. The input must at the most remain equal, if not be decreased; and the price per output unit must at the most remain equal, but ideally be decreased.

In the West technical development was and is pursued to increase output with a decreasing amount of labor. In the Third World, however, the situation is different; capital is expensive and labor is relatively cheap. Hence the preference in those countries for labor-intensive solutions; a preference that logically follows from the above theory.

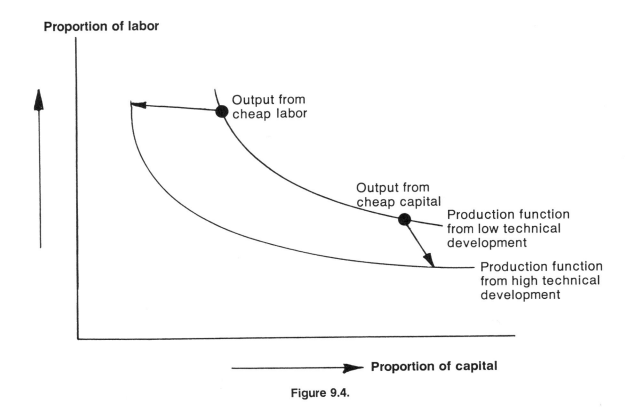

Figure 9.4.

In the Third World, also, the issue is to increase output with an input that remains equal or is decreased. Here again the choice of technique is determined by the costs of capital and labor. Technical progress is, in contrast to the West, aimed at a labor-intensive development. We can summarize this in the diagram shown in figure 9.4.

Traditional and Conventional Technologies

I have said before that the realization of this theoretical possibility is obstructed by the absence of a labor-intensive technology. But even if this technology were present, the choice would be more complex than the theory suggests. In Indonesia lime is burned in kilns of a classical model, the so-called Roman kiln. A modern alternative is the forced-draft kiln, such as they are used at present in Europe. One would suppose that this is an example of appropriate technology, for the former technique is typically labor intensive and the latter is capital intensive. One could therefore expect that in Indonesia the former solution would be preferred. But in comparing the two possibilities we have to make more than one consideration:

We can see that the capital costs are higher in the second case. But the difference doesn't stop at this, for the owners of the Roman kilns can't raise the large sum of money for the forced-draft kiln; it can only be obtained from the capital market. And we see that it is more profitable to invest in the modern alternative for the profit of this investment is 18 percent higher. In the second case, also, the costs of labor are much lower, which is as we expected, and the laborer can earn more. But the unskilled laborers cannot do the new work, which demands technical specialization, so employment is reduced in two ways. The labor income of the region where the kilns are situated—near the limestone mountains—would fall considerably.

The old kilns burn wood, found locally, resulting in heavy deforestation. To change to oil would mean higher energy costs. The modern kiln, however, uses electricity in conjunction with oil. Because there is no electricity available at the place of settlement itself, diesel generators are installed. Therefore the energy costs of the second solution are much higher.

For the consumer the second solution is more attractive because the price of the product is lower. For the time being we can leave out the further difference that will arise when the production is intensified, such as the need for mechanical digging of the limestone, mechanical transport, the greater burden on the roads, and so on.

From the example it is clear that the modern solution will be preferred because of the lower production costs, the higher wages, and the higher profit. This preference means that the production of lime becomes ruled by economic parameters, price competition will cause a lowering of standards of maintenance and wages of the traditional process, and there will be an attempt to lower the price of the product by lowering its quality or adulterating it. But these measures will only quicken the demolition process. As soon as an alternative from the Western economy becomes available, the labor-intensive solution is doomed to disappear.

Although, as Professor J. J. Boeke has already established, the poor economy is often a well-knit structure which cannot easily be broken through, it can be affected by the growth of the rich economy. Where comparable products are made, it is the rich economy that establishes the norms for quality and price. This means that wages can not increase but rather tend to decline.

	Five Roman kilns at 20-ton output per day	One forced-draft kiln at 100-ton output per day
Capital	50,000,000 ..... Rupees ..	390,000,000
Capital costs	5,000,000 ..... Rupees ..	46,000,000
Labor	50 men	6 men
Labor costs	108,000,000 ..... Rupees ..	14,400,000
Labor wages per day	300 ..... Rupees ..	1,000
Energy	wood, oil	oil, electricity
Energy costs	28,555,000 ..... Rupees ..	52,470,000
Material	limestone	limestone
Material costs	4,680,000 ..... Rupees ..	4,680,000
Production	100 ton per day	100 tons per day
Price per ton	5,670 ..... Rupees ..	4,500

Figure 9.5.

We will have to come to the conclusion that it is time to investigate how much the poor economy has weakened, and to take steps to repair this. And with this, socially appropriate technology acquires its most important meaning: all activities which are necessary in the given circumstances to bring the technological network of the poor economy to a proper level of operation. Although socially appropriate technology is a concept which doesn't necessarily have to apply just to developing countries, it is the urgency of their problems which forces us to give them highest priority.

Conventional and Socially Appropriate Technologies

From the foregoing it is apparent that we are dealing with different concepts that can be brought back to distinctive theories.

- Conventional technology can, in the framework of development economics, be seen as that design activity which leads to solutions for production systems which under the circumstances are optimum. In short, where there is surplus of labor and shortage of capital, labor-intensive solutions must be looked for.

- Socially appropriate technology concerns all activities which are directed at bringing the functioning in, and of, the poor economy's network to the desired level, in terms of quantity as well as quality, of the production systems as well as the people who must live with, and on, it.

Socially appropriate technology specifically concerns the Third World, the existence of the poor people in the developing countries. With conventional technology we primarily think of solving the practical problem within a Western network of production and consumption: How can I produce under circumstances that differ from those in the West? In short, conventional technology concerns a production approach and socially appropriate technology concerns an existence approach.

I have said before that it is unavoidable with this approach and with this subject not to switch from quantitative to qualitative observations. I would emphasize the importance of being fully aware what terms we are thinking in, what it is we actually want to do. Both approaches have the right to exist; both are important for the development of the Third World. The biggest difference in my opinion lies in the urgency they give to the problems; those that concern poor economies certainly deserve our closest attention at the moment. And then there is the difference in working methods. With conventional technology it is still possible to undertake a large part of the development process in Western countries; with socially appropriate technology the actual work must be done on the spot, in cooperation with the people that are primarily concerned.

The important task for the West is to prepare alternatives, to compile a catalog of technologies that people in the field can work with. The West also has the task of maintaining contacts to support those institutes in developing countries that are engaged in the development and dissemination of socially appropriate technology. Finally, there is ample opportunity for us to provide training for those who will go to work in developing countries. For it's not only technology—the hardware—that counts; it's also a matter of acquiring insight into the nature of social systems, the ability to work with people of an entirely different culture, and the ability to live in circumstances that differ strongly from those of the West.

Need for Alternatives

From the foregoing it must be apparent that it isn't possible to work out solutions in the West for technical problems in the poor economy of the developing countries. To be able to do that we would need to know much more, and research in that direction has scarcely begun. But we will have to make a start for two reasons. First, we must prepare ourselves for a way of thinking for which we have yet had no training. We must try to think ourselves into the situation of the poor economy, and starting from there we must try to apply our technical knowledge and creativity. Secondly we must collect the data important for finding solutions which could be of practical use. These solutions could be characterized as low-intensity operations, on a smaller scale, demanding less capital, being less one sided and demanding of the work force, using less material and energy.

When we look at poor economies, we find that this type of solution is quite normal there, in all aspects of existence such as food, clothing, shelter, health care, education, entertainment, transport, and communication. The poor economy too has its design code and criteria for choosing from alternative solutions. But as we stated before, it doesn't know a systematic, institutionalized improvement process. And when disturbances are introduced from outside it can happen that working methods that originally were amply sufficient for the subsistence of the people are increasingly overtaxed and finally degenerate. The methods are adjusted, but this adjustment always means a deterioration as has been shown in the example of the lime kilns. Traditional systems can only react by reducing wages, refraining from maintenance, and trying to keep things going by reducing the quality. The poor economy doesn't show any sign of evolution; it involutes. More and more people have to live on this relatively narrow basis, having their means of existence even further reduced by this deterioration in quality. One could ask oneself whether it is a sensible thing to strengthen a poor economy, and moreover, is it at all possible to help the poor? For in spite of everything they are greatly attached to their current ways of living, and they are right because now they at least know what they have. They have seen many examples of improvements that have failed, and they have learned to trust in this structure of existence since few get a chance to benefit permanently from the richer economy.

A general answer to these questions cannot be given. China shows us that it can be done, for the first time in that nation's history the rural people are able to lead a carefree existence. What we don't know is whether the Chinese experiment could be repeated elsewhere. And, if it is at all possible, it can't be said with certainty that the best approach is the one of socially appropriate technology. It remains a matter of choice. It is obvious, however, that there can be no question of being for or against any technology as long as there is no alternative. The efforts undertaken in the West are therefore designed to give content to the wish of many to have an alternative to modern conventional technology. Apart from this, these efforts are of the utmost importance for the existence of the people in the poor economies of the world— people who certainly can't be helped in the short term by more complex approaches. Socially appropriate technology is then at least an intermediate phase, the bridging of a gap.

A Catalog of Alternatives

In the search for technical solutions our main concern is to find answers to the questions: what, for whom, with what? Naturally we can't give general answers, but it is possible to divide each answer into certain categories.

For instance, energy supply for the provision of clothing for a village community can't be separated from the whole of activities around the provision of clothing for that community. The classification only serves to arrange a number of specific fields over the whole spectrum of hardware in socially appropriate technology. And it's important to note that the number of distinguishable fields is very large; a lot of work will have to be done before we can really complete a catalog of realizable alternatives.

We must realize that it won't be easy to fill in this catalog. For even if the special problems that occur in the selection and implementation of the hardware are left out, it is still difficult to find suitable alternatives. Suitable means: possibly usable in a poor economy. This type of solution was characterized earlier as operations of low intensity. In order to find this kind of solution we must establish a new connection between theory and practice.

Consider the development of small water turbines in Indonesia. In the West the small water engine originally played a leading part in the energy supply of small villages and craft industries. But their technical development practically came to a standstill in the last century when all attention was drawn by the big turbines. At the moment most of our knowledge is concentrated on the design and manufacture of large hydroelectric units. In a country like Indonesia, however, there are many minor streams that could be used to supply nearby villages. In order to realize

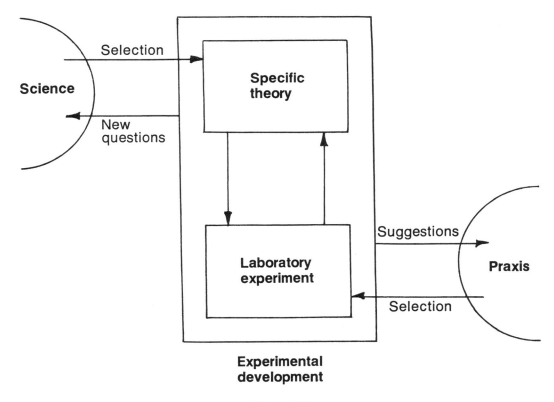

Figure 9.6.

this a small water engine is necessary; so the development that has been stagnant for so many years must be put in hand again.

A Research Method

To accomplish this an interesting working method has been chosen, which resembles the approach of industrial development laboratories, and in which both scientific knowledge and practical experience are used. This experimental development derives its theory from the store of scientific knowledge and tests this theory in trial constructions that are derived from practice. In a diagram it would look like figure 9.6.

This theory is further modified by experiment, and practical suggestions for trying out certain changes can be made based on these experiments. Development work thus clearly interacts with science as well as praxis, and thereby keeps its own identity. The research may give rise to questions that either concern science or praxis; experimental development therefore cannot stand on its own but is always dependent on both science and praxis. The project is a synthesis—a design that can be understood with scientific knowledge and that at the same time fulfills the conditions of praxis. It is my conviction that a useful result cannot be obtained unless one takes up this intermediate position. A development laboratory provides only sterile solutions when the connections with praxis are missing. Concretely, it means that sound fundamental knowledge must be coupled to wide experience in the field of research, for both are necessary to focus a creative interaction process on suitable solutions. And this cannot be done by the isolated scientist.

Conclusion

I have already suggested that there are only limited tasks for universities of technology in the development of socially appropriate technology. They include filling in at least some of the details of the catalog, and supporting those who play a part in implementation. These tasks originate from the special nature of institutions of scientific and technical education. In general, however, I think it reasonable to ask that such institutions should show a sense of responsibility for poor economies in the Third World. They may do this by helping to stop activities that further impair the poor economy, or at least by showing their students to what extent the Western network is capable of interfering with the existence of the people, without necessarily producing any change for the better. So in addition to carrying out direct activities, these institutions must also work indirectly. Contributing to the establishment of socially appropriate technology therefore means omitting certain things as much as doing others.

Chapter Ten

Industrial Liaison

by Paul R. Lofthouse

Although my subject is production systems for use in developing countries, I want to start by saying that there is really no such thing. Every developing country has different problems of its own and each industry, or proposal for setting up an industry, has to be dealt with according to the market conditions and requirements in the country concerned. To explain this I shall discuss a fairly wide range of simple case studies, bringing out the sort of problems that we find ourselves faced with when we try to help developing countries to set up small industries.

First, let's take the setting up of a factory for the manufacture of metal windows. Early in 1972 ITDG was asked as a group to help set up a small-scale factory for this purpose in the Kaduna region of northern Nigeria. The alternative of transporting windows from Lagos, which is 600 miles away, was not favored because of the transport conditions and the risk of loss en route. They felt it would be very much easier if they made their own, from their own extruded materials, even though the level of technology in Kaduna is not high. Extruded aluminum section seemed the best material since it obviates the necessity for painting or other weather protection, because once aluminum has been put up it oxidizes and protects itself.

Window Manufacture

The Nigerians wanted a factory to produce about $150,000 worth of windows per annum. They were prepared to pay for know-how, and they were prepared to pay for people to teach them how to do it. So when I came back to England I contacted all the window manufacturers in England—there are 53 of them at the moment. Out of that 53, only two were prepared to cooperate, in spite of the fact that the "know-how" fee was going to be $20,000.

The operation was organized so that, in the first stage, Nigeria would send to England a foreman, and possibly two men from the assembly shop. The English company would provide the windows at first in a disassembled condition and would also send to Nigeria a man to teach their assembly. Subsequently, the next foreman who would go over would be in charge of the machine shop, and eventually, they would have a complete factory.

Standardization

Unfortunately, or perhaps fortunately, I fell out with Nigeria, because in the course of my research into English companies, I had found that a lot of English companies were going bankrupt because of architects. Every architect who designs a building has a different idea of what a window should look like, and a different idea of the construction of the windows, which means, of course, different extruded sections. And a lot of these comparatively small companies were holding up to one million extruded sections in stock to try and satisfy these architects. And I pointed out to the Nigerians that they simply could not afford to do likewise. But because most of the major buildings are built by the government's Ministry of Works, I suggested that we should agree with the Minister of Works that all government buildings should have a standard type of window, to be used on a modular principle so that one, two, three, four, or five of the units could be employed as required. Any variations would be based on this window, so that you could use the same sort of extruded sections. This seemed to be a fairly rational idea.

I also refused to give them any further information until they had done some market research, because one gets the attitude very often in the developing countries that "I know there is a market, a huge market, for this." But "huge" is never defined. When they had done a proper market survey and reached agreement with the Ministry of Works to accept a standard modular window, then I would proceed with finding a collaborator to make windows in northern Nigeria.

That was in 1972. I had actually filed the papers in the dead file, but exactly two years later I had a letter from the Ministry of Trades and Industries in Kaduna: "Dear Mr. Lofthouse, referring to our discussion on windows, we have now done a market survey; herewith the information. We have agreed with the Ministry of Works, etc. We should like to start up the window factory." Unfortunately by that time the only two companies in England that had been interested had been taken over by a large group and were no longer prepared to cooperate. The end of the tale, however, is really quite satisfactory. By a coincidence, Swiss Aluminum is setting up an aluminum extrusion factory in Lagos, and it also owns Allamarco, which is a window-manufacturing company in Lagos. But it is interested in the North, and Swiss Aluminum is now teaching them how to set up a factory.

This factory will basically be very labor intensive. The total equipment cost of the factory is in the region of $30,000 to $40,000 and the output will be about $2,000,000 a year. Most of it is labor inten-sive. It uses simple machines—circular saws and a little bit of welding equipment, for example. The factory will produce sufficient windows for the whole of northern Nigeria. I am satisfied in my own mind that with the development of the North there is a market for it, although I felt market research figures they gave me were overoptimistic.

Packaging Plant

Now I want to go into a bit more detail on the second example because we feel that this is a case that has actually proved in practice the concept of intermediate technology. Some five years ago, Dr. E. F. Schumacher, the ITDG chairman, and one of my codirectors Julia Porter, were in Zambia doing a general rural survey at the invitation of President Kenneth Kaunda. They were asked whether it was possible to make egg trays. By egg trays I mean the things about 30 centimeters square that carry 30 eggs and are used for transporting eggs from the producer to the packer.

On the strength of this interest from Zambia the group became committed to finding a production process suited to the Zambian requirements. At that time the smallest egg-tray machine that was manufactured and commercially available cost about $300,000 and produced 1,000,000 egg trays a month. But our researches indicated that Zambia wanted only 1,000,000 a year. So it became increasingly obvious that we would have to design a completely new machine. Thomas Kuby, a fully qualified industrial designer, undertook this task and his first reaction was to visit the established egg-tray industry. In fact, there are just two firms that dominate the world market and the one that we approached, said: "Look, we know you want a small machine for making egg trays, but we have been making egg trays for 30 years, and we can tell you that our best

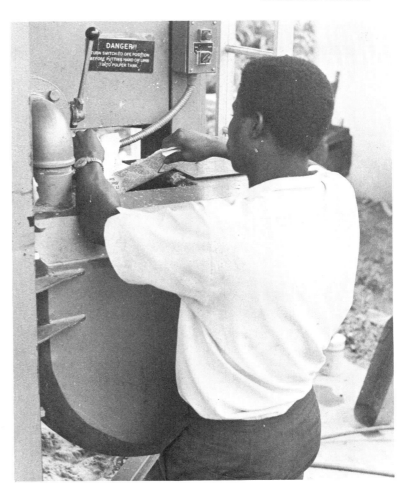

Photo 10.1. Paper pulp packaging unit—shredding newspaper into the pulper.

brains can't do it. It is impossible to produce such a machine." Well, being something of an awkward type, and also having a true flair for seeing engineering in its proper context, Thomas thought that it couldn't be that difficult. So he started from scratch experimenting with making pulp from newspaper in the kitchen. From there he borrowed some facilities at Manchester University to see how best to form the shapes and eventually decided that the only feasible way was with a vacuum that drained the excess water off the pulp through a perforated mold.

And then of course we got really involved. Thomas's design led to a machine for making one-third million—the annual production indicated by our market

survey. In Zambia there were three main egg-producing areas, each with a forecast requirement of about one-third million egg trays a year; this figure had become the target output.

Pulp Forming

This Mark I design incorporated the basic processes used in the conventional plant, even if the size and complexity of the component parts had been critically reexamined. Basically the machine consists of a pulper, a vacuum pump, and a compressor, together with mixing tanks, etc. The raw material is newspaper which is shown in photo 10.1 being fed into the pulper. The pulper produces a four percent

Photo 10.2. Egg trays are lifted from the mold onto the dryer.

Photo 10.3. Finished trays leaving the dryer.

suspension of paper fiber in water through the use of hydraulic pressure that breaks the paper up but damages the fibers of paper to the minimum amount. These paper fibers suspended in water are supplied to a forming station. This was one of the most difficult problems because, to make a mold for an egg tray, is an extremely difficult technological problem. If you look at any egg tray you will see that it has a number of planes over which the paper fibers have to spread evenly. The egg tray is formed by dipping a perforated mold of the correct shape into the pulp, applying a vacuum to build up a layer of fibers over the mold's surface and withdrawing the mold from the pulp so that excess water will drain through the perforations to be recycled. This gives you a fragile egg tray that still contains about 70 percent water and that has to be removed from the mold for drying. Consequently, the mold has to be sufficiently stable to stand a vacuum of 500 millimeters of mercury, which means that it has to be fairly strong. It needs a backing plate and its surface has to be made of fine mesh.

The molds are made, or have been so far, by hand in sections; they are literally hammered on to a former with a little rubber hammer until they're the right shape. The molds are then silver-soldered together and put under a 5,000-ton press for the final forming. You can't press the things to start with because you must use perforated metal, and if you press perforated metal you just distort the holes, and the pulp will go straight through. Photo 10.2 shows a general arrangement of the various components, and shows wholesale egg trays, as opposed to the domestic "six pack" being removed from the dryer in photo 10.3.

The egg tray plant in Nigeria has now been working for 16 months, 16 hours a day, 6 days a week. It has so far produced 700,000 egg trays and they have not had to

replace anything except two bearing seals that were beginning to leak water.

Development

As a result of this installation, we started to get inquiries and it became obvious that the very small plant was very useful for the small country with small egg production. But we were also getting inquiries for a larger plant. So we developed the Mark II. The Mark I plant produces 125 egg trays an hour which works out to one-third million a year. The Mark II plant is basically the same, except that it has two forming stations which are linked together. We have modified it slightly. Instead of hand-operation of the forming station, we are now using an air cylinder to save the physical effort on the part of the operator. This does 250 egg trays an hour. Then we got inquiries from places like Panama and the Middle East who wanted a bigger one. So again we did not redesign; we multiplied again. Four of the Mark II units are equivalent to 1,000 an hour. So we have four sets of Mark II forming stations and instead of using the original pulp-forming station, we have two larger ones and two larger dryers. And that is where we get a certain amount of what the economists persist in calling the "economy of scale." This then, is the Mark III machine which is capable of a 1,000 an hour.

From the group's point of view, the interesting result of this is that we were approached by the second of the large egg tray firms who said, "We are currently turning down a lot of orders for plants which are smaller than we can make, or know how to make, but which are just the right size for your Mark III." Now ITDG has no money and is noncapitalized. The big plant will probably retail in the region of $200,000, which is still economically very viable. But we haven't got the sort of

Specifications of the Paper-Pulp Packaging Unit

The units are designed to produce items of paper pulp formed for packaging—with a maximum unit size of 360 millimeters square x 100 millimeters deep—using waste-paper, printers' off-cuts, newspaper, etc.

EACH COMPLETE PLANT COMPRISES THE FOLLOWING ITEMS:

"Mk 1a Plant"
One Pulp Preparation Station
One Vacuum-forming and Transfer Station
   Mechanically Operated
One Tunnel Dryer

"Mk 11 Plant"
One Pulp Preparation Station
Two Vacuum-forming and Transfer Stations
   Mechanically Operated Together
One Tunnel Dryer

LABOR REQUIRED:

Three Operators, Plus Supervisor

OUTPUT:

"Mk 1a Plant"

120–130 Units per Hour

"Mk 11 Plant"

200–230 Units per Hour

OPERATIONAL MATERIAL AND POWER REQUIREMENTS PER HOUR:

"Mk 1a Plant"
100 grams Soluble Wax
100 grams Aluminum Sulfate
10 liters Water
8 kilograms Paper

"Mk 11 Plant"
200 grams Soluble Wax
200 grams Aluminum Sulfate
20 liters Water
14 kilograms Paper

Electricity:
9.5 kilowatts for Pulp Preparation and
   Forming Station
35 kilowatts for Dryer

Electricity:
19 kilowatts for Pulp Preparation and
   Forming Station
66 kilowatts for Dryer

NOTE: Dryer can also be heated by steam, oil, or gas.

SITE REQUIREMENTS:

Covered Area
Mk 1a 6 meters x 12 meters

Mk 11 8 meters x 15 meters

PLANT SIZES AND WEIGHTS:

Pulp Preparation Station —3 meters high x 2 meters wide x 2 meters deep—Weight 3,000 kilograms
Mk 1a Forming Station —2.2 meters high x 1.4 meters wide x 1.2 meters deep—Weight 750 kilograms
Mk 11 Forming Station —2.2 meters high x 3 meters wide x 1.2 meters deep—Weight 1,500 kilograms
Mk 1a Dryer —9 meters long x 2 meters wide x 2 meters high—Weight 1,800 kilograms
Mk 11 Dryer —9 meters long x 3 meters wide x 2 meters high—Weight 2,500 kilograms

ALSO AVAILABLE:

Mk 111 Plant Giving 1,000 Units Per Hour—Details on Request.

Table 1.

money to market a plant like that and carry the capital cost over until we get the money back. So I have signed an agreement with this firm whereby they will buy our Mark III plant to fill the gap in their product range that, after 30 years, they haven't been able to discover a successful solution for.

So the group now feels that it has intermediate technology in its true sense—the smallest, the middle, and the larger—every step right up until the large mass-production plant takes over. We naturally get something out of it. We make a profit on the sale of the plants to this company which goes back into the group to help its main work. Consequently we feel that we have proved without any reasonable doubt that there is quite a lot of virtue in designing a small plant. If a world monopolist ends up by asking to buy it, there must be a real need for this scale of production.

Technical Inquiries

I propose now to go from the sublime to the ridiculous. In the course of the year, any year, we answer many technical inquiries. They are of all types, from very small to very large. I want to discuss what is probably the smallest. I had a delightful little letter about 12 months ago from a man living in the middle of Africa. The letter said: "Dear Mr. Lofthouse, my business is providing sand for a cement works. My men walk down a slope 130 feet long to the riverbed. They fill a basket with sand and they walk up and dump it on the top of the slope." And he sent two little tiny photographs to actually show me the actual river. "I have recently been able to get hold of a five-horsepower diesel engine. How can I use this to improve my production? I have $20."

I wrote to him and said: "$20 is not very much money. I would suggest you go to a dump, and find a Land Rover and buy the power-takeoff unit cheap. Put a belt on that and a pulley on your diesel. Beg, borrow, scrounge, or if necessary, buy about 140 feet of ⅜-inch wire rope. Get a large bucket and have two wheels welded on it and drill holes in the bucket. Your men can carry the bucket down the slope, fill it with sand, and you can then tow it up using the Land Rover winch with your diesel engine."

Well, I never expected to hear anymore, and so I was rather delighted about six months later when he sent a letter which said: "Dear Mr. Lofthouse, I thank you very much, I have doubled my production." I am perfectly sure that in about another 18 months he will probably write again, saying, "I now have $100, how can I improve my production further." Then I will show him how to make a conveyor. Or something of that nature.

Intermediate Technology Workshop

Perhaps these interesting examples, explaining the practical application of intermediate technology, show why I maintain that one cannot talk about any specific type of production. I now want to go on to a little factory which actually started this industrial liaison exercise in the group. Originally this particular exercise was founded by the Ministry of Overseas Development as a research project. The original conception of the research was to have a man in Africa to research into whether Africans wanted small-scale industry; and to have myself in England to research whether the United Kingdom, or the Western world generally, could provide suitable small-scale equipment.

Not being an academic or a research person, I went over to northern Nigeria to see Will Eaves, the man there and I said, "We've got a three-year research program and instead of walking round and asking everybody if they want a small-scale factory, let's set one up and see whether it works." So that is precisely what we did. We set up a factory and the Ministry put in $2,000 and North Central State put in $3,000, and provided us with an old ex-colonial bungalow. They also provided us with 17 primary-school leavers, the sons of subsistence-level farmers who would, in the normal course of events, have had no future and who would probably have ended up as beggars, or something of that nature, because they could get no further.

Hospital Equipment

The first thing we taught these boys was how to reroof the bungalow, wire it and so on, and then we started into production. Now at that stage, Nigeria was desperately short of hospital equipment. Will Eaves was trained at University College Hospital in London as a hospital

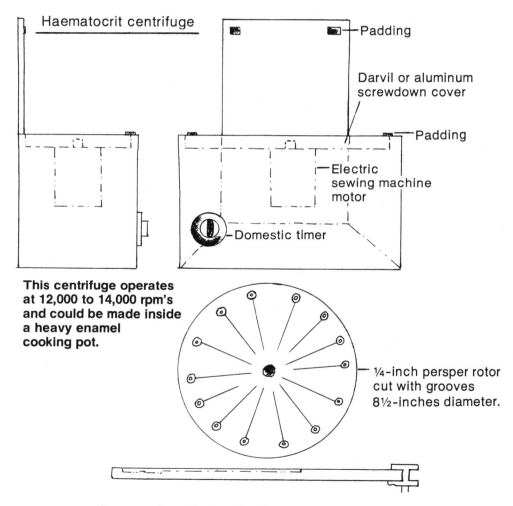

This centrifuge operates at 12,000 to 14,000 rpm's and could be made inside a heavy enamel cooking pot.

¼-inch persper rotor cut with grooves 8½-inches diameter.

Cross section of rotor showing groove and pit cut with circular saw or homemade scraper. The capillary tube lies inside this groove. Depth of groove — 1.5 millimeters.

Figure 10.1. Centrifuge made at the Intermediate Technology Workshop, Zaria, Nigeria.

technician and is himself an innovator almost to the point of being an inventor. The only resources he had available were building materials, such as conduit, pipe, angle, and what he could scrounge in the way of trolley wheels and things of that nature. So he designed the whole range of hospital equipment, for example the centrifuge shown in figure 10.1 to be made from these local raw materials and got orders from the Ministry of Health to equip at least three hospitals. And it was all made by these primary-school leavers who came straight into the works where they were taught to use the only equipment that we had—an oxyacetylene welder and a hand-operated hydraulic bending machine. Except for an electric bench drill, all work was done by hand-tools. This factory has now been running for three years and the original $5,000 that was put in to set it up is still turning over. The boys only stay there a year, but the factory is training 35 boys a year in basic engineering principles and it is costing nobody anything. And the students are filling a very important local requirement.

The factory produces more than hospital equipment. For example, Christmas came and the local children had no toys, so we started to produce swings and rocking horses and things like that. They went like a bomb because they were very cheap and they really were very well accepted. Then, a local firm there was doing some research into tomato growing and wanted special cages to protect the tomatoes. We built the cages. It became a general engineering workshop, with the accent on hospital equipment, but capable of doing any small simple engineering work.

Entrepreneurs

The purpose of this venture was to train these boys to stand on their own feet. Now entrepreneurs are born, not made; at

least that is my opinion. You can send them to as many management courses as you wish, but you cannot make a manager unless he is born with the right instincts. Our first 17 boys immediately got jobs in local industry and went on to the next stage, which was a proper apprentice training to get their qualifications so they could get union rates and all the other things. After about six months two of them became tired of this and got local *alhadjis* to finance them. One of them went to the old city of Zaria, got himself a wooden hut and is currently making a very nice living from sheet-metal work. He makes all types of sheet-metal work, particularly office equipment such as filing cabinets. The other one set up in competition with the main organizations and is also doing very well.

Wheelchairs

One of the most successful things that developed in the workshop was a wheelchair. There are in the north of Nigeria a lot of disabled people, many of whom have been crippled in cattle stampedes. The Muslim religion is such that if a man is disabled, it is an act of God, and his fellow-people are supposed to look after him. Nevertheless, the invalids don't feel very happy about it. For this reason we developed a very simple wheelchair, basically constructed of building materials, a bit of plywood and a couple of bicycle wheels. The front wheel is one of the heavy wheels that you find on a delivery boy's bicycle. It is chain driven by a hand crank.

Now the beggars have a union, of course, and the head of the beggars' union came to Will Eaves one day and said he would like one of these wheelchairs. So Will told him that they normally sell for $30—it costs us $24 to make them—but he could have one for $24. But he didn't have $24, and so he said, "We will save up. We will have one wheelchair between us so that at least one of us can be mobile." As he was painfully dragging his smashed legs down the drive, Will asked him why they didn't do some work. "We're disabled. Most of us have got no legs." to which Will replied, "You've got hands." Accordingly he bought a number of fretsaws and scrounged (a lot of our work is done through scrounging) off-cuts of plywood from the local factories and taught the beggars how to use a fretsaw. They started off by cutting out and selling the letters of the alphabet, then the emblem of the Northern cross, which is very popular there, and then they went on to jigsaw puzzles.

There are now 10 beggars in the group. They are completely self-supporting, and have now purchased for themselves five wheelchairs which they take in turns to use. It is really frightening to see these fellows. I hope the wheelchairs are as stable as they look, because they go tearing madly along the road at about 10 miles an hour and they have no idea of how to take a corner, but they get round somehow!

But it is extraordinary to see the change in those beggars over the last couple of years. They are completely rehabilitated now. They are well dressed; they are happy, and are completely different people. It also has had a kind of a spin-off in that when I go up there now, no beggar ever asks for alms. It just shows that if one has the imagination, almost anything can be done. I think the total cost to the group, which as a matter of fact came out of Will's and my pockets, was $10, and there are now about 10 beggars earning their livelihood. The thing that particularly surprised me was that the local leper colony wanted to do work for itself. And the disabled beggars welcomed the lepers into their circle and taught them how to do it. For an African community to welcome lepers into their circle and teach them a skill was, we felt, quite a success.

Local Design

I was talking about this about 12 months ago when I was at Strathclyde University and an economist said, "That's all very well but that was the best they could get. If only they had had decent wheelchairs, how much happier they would have been." But on that occasion, fortunately, I had the answer, because while I was there, one of the rich beggars (there really are some rich beggars who do make some money) came up with an imported wheelchair of which the delivered cost at Zaria was $100. And he asked Will if he would take it in exchange for one of his. Will refused because it was utterly useless, absolutely useless for the conditions there. It would have been marvelous in this country where you have smooth surfaces, but to try and use a Western wheelchair on those roads would just not have worked. The same applied to the hospital equipment, which is why we redesigned all that. It just does not work unless you know the conditions.

Small Industry in Mauritius

I think we will now take a different sort of problem. Mauritius has plenty of educated people but few jobs for them. Being educated they can also think rather well, and there is a very explosive situation developing. Mauritius wanted to know how to employ these people.

They have an organization for the lesser-trained people who work on government schemes, such as construction. But it is the secondary school children leaving school who are becoming a problem, and Mauritius would very much like to set up industries which can employ them. Now the only really interesting industry that I saw there was jewel drilling. The Swiss watchmakers turn jewels to size and send them to Mauritius where the hole is drilled through them. This is a labor-intensive and highly skilled job, and because there is very high value and very low volume, you can afford the transport. Unfortunately the procedure was changed. The jewels are now drilled with lasers, and this is being done in Switzerland. However, most of the jewels, as well as being drilled, have to be countersunk. You can't countersink with lasers so they have altered the machinery and are now countersinking the jewels in Mauritius.

This example parallels a lot of requests that we are getting in increasing numbers from all over the world, particularly with reference to electronic equipment and other electrical equipment in general. A lot of people are saying that they have comparatively cheap labor and can they not import the components for the small batch-produced stuff that is not mass-produced, build the electronic equipment, and ship it back? Now I must admit I have not had much success with this so far. Also, I am worried from the personal point. I may be unduly pessimistic but one is watching Hongkong and places like that rather carefully at the moment. For years these free-port areas have had very good business because they have cheap labor and can compete because of their cheap labor. With the coming of the mass media and the general growth of knowledge all across the world, most of this labor is now becoming organized, and I have a horrible thought in the back of my mind that each of these places, probably starting with Hongkong, is going to end up as a wilderness. Once the labor force gets what we would call a reasonable return for its labor, it will no longer be an economic proposition to send the stuff over and rising transport costs will accelerate this trend. I can see a lot of these cheap labor areas being much worse than they were when they started.

Hand-Tools in Ghana

I want now to take another example that shows that, where you have a farsighted management, there is quite a lot that can be done. This was in Ghana, in Kumasi University's Technology Consultancy Center which asked me if it could make hand-tools, particularly chisels, screwdrivers, and hammers. And I knew perfectly well what the potential market was. I said that they couldn't do this because they would need different metals for chisels, and different metals for screwdrivers. In addition, by the time they had gotten the metallurgical control, the heat treatment, the drop-stamps and all the other things they would need for manufacturing this sort of product, it would not have been economical to do it using the local blacksmiths. The capital cost is such that you could produce enough in a week to last Ghana for a year. But nevertheless I asked them to leave it with me and I went to the Small Tool Trades Association in Sheffield and talked it over with them. They suggested that I go and see Footprint Tools, the well-known tool manufacturers in England. I asked why they suggested this particular firm and they replied it was because it did not have an accountant on the board! So I went to see them and the result is that Footprint is exporting to Ghana about 10,000 heat-treated, finished blanks a year. The people in Ghana are grinding them, edging them, polishing them, putting handles on them and selling them as Footprint tools. Footprint sold them two high-speed wood-turning lathes to make the handles. Footprint has not been entirely without benefit from this because they themselves are desperately short of hardwood handles which they are now importing from Ghana. The managing director told me, "We will make no money out of this, but once Ghana realizes that Footprint tools are good tools, when they want to import the more sophisticated tools which they cannot make themselves, they will ask for Footprint, and with any luck, in 10 years' time we should have a developing market in Ghana."

Exports

This attitude is something that I welcome very much in a large company—to see that they are prepared to look 10 years ahead, to build the name of the company up, to build the reputation up, so that in the future they will have a developing market. I think I should mention here that I do get criticism from time to time from various organizations in England, who say that I am just "cutting their throats" by teaching these people to make their own goods. I am ruining their export market. But this is not strictly true. Because if one can build up the standard of living of the developing countries, potentially one is really building up a completely new export market. When they have an increased standard of living and some free cash, they will be importing things from us. It will certainly change our export market completely but, provided we are prepared and lively enough to change with it, we will be able to import goods from them that they can produce better or cheaper. In turn they will be a potential export market for goods that they cannot produce. There are a lot of items that we produce on which they could not possibly attempt to compete with us because of the small market potential that they have. It is a completely changing picture, but I don't think necessarily that by helping the developing countries to develop we are cutting our own throats. In fact, I think myself that we are investing in the future.

Glassmaking

I want now to touch on an interesting possibility for the future which started from inquiries that we had from all over the world concerning glass. I was horrified to get a letter from Chile, saying, "Please can you find us a supplier of glass jam jars. For years we have been importing jam jars from the United Kingdom to put our jam in."

The mind boggles at the thought of boatloads of empty jam jars being transported there to be filled with jam and then exported back again to the Western world. I thought this was absolutely crazy. At the moment glass production, certainly in England, is very limited, and the reason they wrote to me was that they could no longer import the empty jam jars. So I started looking into this.

I talked to various people, all of them very knowledgeable about glass; among them was the Chairman of the Glass Research Association. I said, "I have done a little bit of homework and as far as I know glass is basically silica sand, limestone, and soda ash with a few trace elements like arsenic to clear it. So what is the problem in making glass?" He said that the problem is soda ash, which is only produced by ICI and without soda ash you cannot make good glass. I was feeling just a little bit naughty that day, so I said, "How long has ICI been running, 50 years, 60 years? Could you then tell me how the Egyptians and the Chinese made glass before ICI was ever heard of?" This caused a rather deadly hush, I'm afraid. So I consulted a chemist and found that soda ash is only used as a flux and is easily obtainable from either seaweed or wood ash, and in fact, if you just add seaweed to the melt you get a slightly green glass which is rather attractive—you don't have to use soda ash at all.

We now have a plant under design of a similar scale to the egg tray plant I discussed earlier, which will probably be sold as a package deal. The question of the furnace is interesting. A furnace normally costs about $40,000 because it needs highly sophisticated refractory materials.

The reason why you need to have highly sophisticated refractory materials is that a furnace is expected to last 10 years. But if you pose the question of what happens if there is local refractory clay and the people are prepared to rebuild every six months, the experts will admit that such an approach is perfectly feasible. They just don't recognize that these people might not mind labor; the fact that they have to rebuild the furnace every six months does not matter. They have got the labor; they have got the clay; and if they can do it themselves, it is cheaper than importing these highly sophisticated pieces of equipment. Although I am only doing the initial costing at the moment, and we haven't yet even built a prototype plant, I am nearly certain that one could build a factory using what is called a "day tank," to make 1,000 jam jars an hour from local raw material, for less than $50,000. A shipload of empty jars going from England to Chile just does not compare economically.

Incidentally, the glass this plant will make will be slightly green and will have little tiny bubbles in it, because it seems pointless to import arsenic, and various other trace elements, which are extremely expensive and are only used to clarify the glass. I was interested to see that major stores are actually selling as a curiosity, bottles made of green glass with little bubbles in them at $3.50 each.

Economic Viability

I just want to end up with one small example because it goes back to the other extreme again. I received a letter from a man called Father Mulligan in Grenada which said: "Dear Mr. Lofthouse, I have a circular sawbench, I have an electric motor. The shaftsize is such and such, the motor horsepower is so and so. Herewith a check for $40, can you please send me some pulleys and belts?" Which of course we did. Now I quoted this case when I was having something of an acrimonious argument with 15 economists who were saying to me, "Why do you answer these inquiries; how do you know that the answer you give is economically viable?" My reply was that if I had written to Father Mulligan and asked if it was really economically viable to set up a sawbench in Grenada, what good would it have done for anybody? However, we have had a lot of letters from Father Mulligan since. I have helped him obtain a lot of information and equipment and at the moment we are actually supplying two 45-horsepower electric motors because he has now set up his own sawmill. But I happened to meet the High Commissioner of Grenada about six months ago, and I asked him if he had ever met Father Mulligan. "Yes, he's a marvellous fellow. He's got the best workshop in the islands. Where he gets his equipment from I don't know, but I wish I did." So I told him.

Chapter Eleven

Some Social Criteria for Appropriate Technology

by Ton de Wilde

In an article for the Gandhian Institute of Studies in 1961, E. F. Schumacher wrote: "It seems obvious that higher productivity is better than lower, and that the highest is the best. Modern technology, as developed mainly in the West, appears, therefore, as the most desirable object for adoption by poor countries, which are poor because their productivity is low. It is often suggested that underdeveloped countries have even a special advantage in being latecomers, as they can 'jump' over the intermediate stages of technological development, which have disfigured life in the rich countries during the nineteenth century and can now—with aid from the wealthy—go straight to the highest level of technology to produce affluence without exploitation.

"This idea of the great 'jump'—from bullock cart to jet engine, so to speak—is often advanced to mollify those who are afraid of machine civilization as starting the full development of individual man, destroying his dignity, and enslaving him to mechanical forces. Modern technology, it is suggested, can produce so much wealth with so little effort that men will be set free for the first time in human history: even if their 'work' becomes meaningless, they can 'live' during their leisure hours. Industrialization on the level of nineteenth-century technology may have been a dark tunnel; but twentieth-century technology leads straight into the light of affluence."

Growth

In the context of a socially appropriate technology, I want to examine the way technology is used in society, and then look at technologies as implemented by several organizations other than ITDG. First then, the way technology is used in developed countries, such as the Netherlands. In fact, I can only pose questions— questions about the appropriateness of this twentieth-century industrial technology that should "set [people] free for the first time in human history." For example, when I go to a car factory, I would see that all its modern transfer lines and automatically controlled machines are not working for most of the week, and there are people there for only half the week. Mass-production technology is very sensitive to market fluctuations; nevertheless, I might understandably get the impression that we are trying to produce things without people. Or in the cases where it is not yet possible to produce goods without human beings, that we are trying to design our factories so that the worker only has to move his right arm up and down and push his left arm forward two inches, twenty times a minute, eight hours a day.

But I want you to consider if he really has an enjoyable or fulfilling time. For when this man gets older he might be unfit to work anymore. At the moment an employee of over 45 years has a 20 percent chance—one in five—of being one of those who are no longer fit to work. Or he might be unemployed, and in our society this may not take long. With a 10 to 12 percent unemployment rate he can easily join the one in ten people who have no work, not even the assembly-line work I have just described.

Purpose

But here we must ask what our aim is. Is it to develop factories along these lines, to continue pursuing a technology that sets the people free from work? The technologist can only aim for a society where people are consumers and are no longer producers. In our Western rationality, in our dialectic way of thinking, we distinguish between consumers and producers, between life and work, between capital and labor, between materialistic things and spiritual things or feelings. But what happens when the decision makers base their decisions on models that only recognize a part of our human being? Now, on the one hand, we have our cars—sometimes with stereo—our color television, our household gadgets, our parties; on the other hand, we have our psychiatric hospitals. And there are simply not enough of them to cope. We do not have enough psychiatrists and psychologists to help the people who run their heads against the wall—people who want to work but who can only go each week to the social service to be paid for doing nothing. It must be clear by now that to me, technology means the labor situation and the whole technological infrastructure.

The technological network is concerned with a mass of information, the speed of life, places crowded with cars, a lot of garbage in nature, the spoiling of our ecological system. For some people in some situations this system has its benefits, but we must realize that there is also a bad side for other people in other situations. This is true not only at a national, but also at a world level. One part—the 20 percent in the developed countries—is well fed and has some kind of work, while the 80 percent in the less-developed countries are badly fed, and there is no work for 30 to 50 percent of them.

What is the cause of all this? I don't believe that the so-called capitalist factory designer really wants to exploit his laborers, sending them to psychiatric clinics, making them disturbed and unfit to work. These crises are the result of the best humanitarian intentions to replace hard human labor by mechanization, to decrease human suffering, and to increase life expectancy. But if this is our intention, we must be conscious of what our aspirations are, what sort of society our children will have to live in. This means that we have to think about the use of technology in our present society. And perhaps to formulate it in the way that is done in the second Club of Rome report: How can we progress from an undifferentiated, unbalanced growth to organic growth? In the words of Professor Sagata Dasgupta, director of the Gandhian Institute of Studies, we should move back from our growth-oriented technology to a society-oriented technology.

In fact this was the independent conclusion of discussions that have taken place in the Brace Research Institute in Canada. This institute, originally only working on technological solutions to water-supply problems in arid areas, learned through long experience that technology should be society oriented.

In collecting material for a *Handbook on Socially Appropriate Technology* there were some hard discussions about the relationship between society and technology. Frede Hvelplund from the School of Economics and Business Administration in

Aarhus, Denmark, was involved and pointed out that appropriate technology should be seen as a process. He suggested that if it was possible to take a photograph of appropriate technology, the photo might indicate a small-scale factory in action. A static picture will show us many of the technical details which have been characterized as labor intensive, simple, local, etc. If instead, we could take a moving film, we would see a dynamic process, and would notice the relationship between people and technology, between groups of people and technology, between technology and organizations in the local social-economic structure. In trying to describe this process we can distinguish four main components: the resources, the people, the technologies, the economic and political structure. All of these components have to be in basic good shape in order not to constrain the process. The technology can be very good and still the structure can hinder a process, or the technology can be excellent and yet be restrained by people who are culturally deprived.

Criteria

Any of these four components can be regarded as parts of an appropriate technology process and any of them can inhibit such a process or make its performance extremely difficult. For example, if self-esteem, creativity, awareness, or capability are not present in the people, it often makes no difference whether the structure is good and the technology is adequate. If the structures are binding or hindering the creativity of the people, it is also of little use to introduce an appropriate technology. Therefore, Hvelplund has generated 15 criteria which should be kept in mind when starting an appropriate-technology process:

1. The technology organization must be adapted to local cultural and economic conditions.

2. The tools and processes utilized must be under the maintenance and operational control of the local workers.

3. The technology should, wherever possible, use locally available materials.

4. If imported materials and technologies are used, some control of them must be available to the community.

5. An appropriate technology process should, if possible, utilize locally available energy resources.

6. The technology should be flexible in order that the community does not get bound to a wrong direction.

7. Research and action should be integrated and locally carried out.

8. The process must tend to produce items which the majority in a country can currently afford.

9. It must create jobs for all people in society and in this way be able to utilize local human resources. (This point is necessary to ensure that the absence of formal technical education does not hinder a person in getting a job, and that the work is transformed to the cultural patterns of the local workers, instead of trying to conform the local workers to the technology. In the long term it may also prevent the developing country tending to approach the situation of industrial countries, where a constantly increasing fraction of the population—old people, young people, women—is defined as unable to participate in the working and society-creating process.)

10. It must be able to compete or, if that is not possible, avoid the competitive sphere. (In cases where it is impossible to gain any control over the price mechanism

on the international market, or where the local backing power is too low, it might be better for the majority to minimize trade. This minimum trade alternative simultaneously increases the local bargaining power.)

11. Technology must prevent external cultural domination.
12. An appropriate-technology process should be ecologically sound.
13. Any appropriate-technology project that deals with primary products and raw material extraction must also try to establish manufacturing units for this product.
14. The appropriate-technology process should constantly be innovative in order to improve the human and material conditions of the local people through the use of new organizational types and new technological devices.
15. A process should not only be appropriate at a local level but also be formed in such a way that it takes regional, national, and international conditions into consideration.

Bearing these criteria in mind, we should realize that socially appropriate technology implies a choice for a certain way of life. It is a political choice. It is a choice for each individual to make for himself.

International Rice Research Institute

I want now to describe some of the other organizations, apart from the Intermediate Technology Development Group, which are working on "Appropriate Technology." From the examples, you will see that some pay more, some less, attention to the social aspects of appropriate technology.

The first one I want to describe is the International Rice Research Institute. In 1967 the Agricultural Engineering Department of IRRI started a program in the Philippines to develop low-cost, small, power-operated machines to be manufactured in Asia. Amir Khan, the head of this department, analyzed agricultural mechanization technologies and distinguished two approaches. "The Western approach emphasizes dryland farming with large, high-powered equipment. It employs capital-intensive technology and has evolved from a primary emphasis of replacing human labor with machines. Introduction of such a technology in the developing countries tends to create labor surpluses, and is not so desirable in the highly populated tropical Asian region. Many attempts have been made to introduce this technology in the developing world. Nearly 30 years of effort, however, has produced rather insignificant results. In India, where tractors of over 35 h.p. have been introduced since the end of World War II, only one percent of the total arable land is worked with such tractors today. Mechanization in Japan has not followed the Western approach. Rice is a major crop in Japan and is grown on small farm holdings under wetland conditions. The high support price for rice, coupled with the country's rapid industrial growth and a rising standard of living, has resulted in the mechanization of agriculture using relatively low-powered, yet quite sophisticated, farm machinery. This equipment has been developed to meet the requirements of the Japanese farmer but is still far too complex and uneconomic for the rest of tropical Asia."

In his analysis Amir Khan concludes that the "Developments in the industrialized countries are further widening the gap between the capabilities of their mechanization technologies and the needs of the farmers in developing countries. In

163

Photo 11.1. Bellows pump at the International Rice Research Institute.

Western countries there is a trend to use high-powered machines with sophisticated control systems; to some extent the Japanese development is following a similar trend, along with an ever-increasing complexity of their machines. These developments are making it increasingly difficult to introduce imported agricultural machines in the developing regions . . . .

Even the selling price of imported farm equipment in the developing countries is approximately two to four times its price in the country of its origin. Furthermore, this expensive equipment must compete with low-cost local labor . . . . Most developing countries suffer severe balance of payments problems. Even a simple calculation indicates that large-scale importation of

164

Photo 11.2. Seven-horsepower tiller designed at the International Rice Research Institute.

equipment to mechanize tropical agriculture is not possible, because of foreign exchange shortages. This dilemma can be solved through the development of appropriate agricultural machines in line with the needs of both the farmers and the manufacturers in the developing countries.''

It can be seen in table 1, which lists some agricultural mechanization indicators for 11 Asian rice-producing countries, that in Japan a good combination of mechanization and labor is possible. In Japan, mechanization, measured by the power available per hectare, is 2.664 h.p. per hectare which makes it by far the most mechanized country in Asia, while it is second in labor intensity. The agricultural mechanization department of IRRI has tried over the last years to develop appropriate agricultural machinery to meet the need of other Asian countries and to

Photo 11.3. Cretan-type sail windmill irrigating plots of land adjacent to Omo River, Ethiopia.

Some agricultural mechanization indicators for 11 rice-producing countries in Asia.

Country	Arable land per holding (hectare)	Agricultural working population/ hectare	Horsepower per hectare				Horsepower per agricultural worker	Labor hours for rice cultivation/ hectare	Net domestic agricultural production US $	
			Human	Animal	Mechanical	Total			Per person	Per hectare
Ceylon	1.59	1.20	0.120	0.148	0.110	0.378	0.009	Not available	293	352
India	2.62	0.90	0.900	0.204	0.008	0.302	0.009	1,000	148	133
Iran	6.17	0.37	0.037	0.048	0.154	0.239	0.418	Not available	417	154
Japan	1.06	2.16	0.216	0.120	2.664	3.00	1.231	1,400	626	1,350
Korea	0.90	1.96	0.196	0.236	0.003	0.435	0.0013	830	244	477
Nepal	1.22	2.49	0.249	0.480	0.004	0.733	0.0016	Not available	99	236
Pakistan	2.37	1.09	0.109	0.288	0.013	0.410	0.012	Not available	154	169
Philippines	3.66	0.71	0.071	0.104	0.023	0.198	0.030	800	242	186
Taiwan	1.11	1.95	0.195	0.164	0.164	0.505	0.074	1,300	349	696
Thailand	3.64	1.10	0.110	0.184	0.054	0.348	0.050	Not available	102	112
Vietnam	1.57	2.10	0.210	0.244	0.023	0.477	0.004	Not available	203	421

Source: APO Expert Group Meeting on Agricultural Mechanization, APO Project SYP/III/67, Tokyo, October 1968, vol. II.

Table 1.

compete with imported machinery. The designs are developed jointly with local manufacturers and tested for several years before industrialized production is started.

In 1971 they began with the production of a power-tiller. On balance, they feel that this scheme was a success, not only for the farmer but also for the industry where it created an extra 400 workplaces, while the investment cost for each workplace was $99 compared with the normal $1,000. The design of the power-tiller requires only simple manufacturing operations such as cutting, welding, and drilling. For the more complicated components standard designs, already available in the Philippines, were used; for example, normal gear transmission was replaced by a transmission from available motorbikes. Table 2 gives details not only of the power-tiller but also of other agricultural machinery projects.

It is important to note that although IRRI works within some of the criteria I outlined earlier, its projects still lie within the framework of a Western economy. Its main criteria are that all the designs should be commercially viable, and that they should be able to compete with imported machinery. So IRRI does not pay much attention to improvements in the use of animal-drawn equipment.

I have doubts whether only mechanical mechanization should be emphasized in Asian agriculture; I have counted the percentage of human, animal, and mechanical horsepower used per hectare in the 11 countries mentioned in table 1. These show that animal-drawn equipment still provides more than half the total power used per hectare.

Planning Research and Action Division

The Planning Research and Action Division of the Uttar Pradesh State Planning

Institute in India works along similar lines to IRRI in developing designs in collaboration with local manufacturers. M. K. Garg, the head of the Rural Industrialization Department, has introduced what he calls the "technology package," which includes both the hardware, actual machinery, and the software, the training of the laborers and management, the marketing, and the form of ownership. PRAD has been involved in rural industrialization since 1959 and one of its best-known projects has been small-scale sugar factories. The need for crystal sugar in India is quite large; the total consumption a year is about 5 kilograms per head of the population (compared to the Western countries' 50 kilograms per head per year). Compared with installed refining capacity there is also overproduction of sugarcane, which is a crop that is easy to grow. Each production unit of the large-scale vacuum-pan process needs an investment of about $5 million, which is far too high for Indian villages. Over the past several years there has been a decrease in the Khandsari industry, the traditional industry that processes sugarcane to a kind of brown candy. In 1910 this industry processed about 25 percent of the grown sugarcane, while in 1955 it processed not more than 2 percent. The percentage of sugar extracted from the cane when processed into crystal sugar through the vacuum-pan technique is 75 to 80 percent, while the indigenous Khandsari process can only recover 42 to 45 percent of the available sugar. These were just two of the reasons why PRAD started to search for a simple, yet efficient, process. Their first design was produced in 1959 and, after 15 years, more than 800 small-scale sugar plants had been installed.

Table 3 provides statistics in respect of (a) a modern mill of the type now being built with 1,250 tons per day cane-crushing capacity and (b) a standard open-pan sulfitation plant of 80-tons-a-day crushing

IRRI agricultural machinery development project, June 30, 1974.

	Asian Manufacturers' Report										
	1	2	3	4	5	6	7	8	9	10	Total
1. Number of machines manufactured till June 30, 1974:											
Power-tiller	3,900	11	1,234	7	457	—	—	—	—	—	5,609
Batch dryer	—	—	—	35	—	—	4	—	74	—	113
Axial flow thresher	20	—	2	8	20	60	9	9	—	—	108
Table thresher	—	—	—	—	84	—	—	—	—	—	84
Grain cleaner	—	—	—	20	—	—	—	—	—	—	20
Bellows pump	—	—	—	100	—	—	—	—	—	—	100
Multi-hopper seeder	—	—	—	232	—	—	—	—	—	—	232
Single-hopper seeder	—	—	—	338	—	—	—	—	—	—	338
Power weeder	—	—	—	—	—	—	—	—	—	7,500	7,500
2. Production capacity utilization percentage:											
Current	85	NR*	50	95	75	70	100	NR	95	NR	
Before start of IRRI machines	70	NR	20	90	45	50	75	NR	90	NR	
Percentage of change	15	NR	30	5	30	20	25	NR	5	NR	
3. Number of new workers employed	399	4	90	65	42	18	12	5	45	NR	680
4. Additional capital investment (US $)	76,000	NR	120,000	10,000	8,300	1,400	12,000	1,500	3,800	NR	233,000
5. Additional capital investment per worker (US $)	199	NR	1,330	154	197	78	1,000	300	84	NR	330

NR* — not reported by manufacturer
US $ 1 — Philipino $ 6.60

Table 2.

	(a) Modern mill	(b) Open-pan sulfitation plant
Capacity (maximum crushing in tons/day)	1,250	80
Output in an average season (tons of sugar)	12,150	640
Investment required (land, buildings, plant and machinery in millions of Rupees)	28	0.6
Total employment (permanent and seasonal)	900	171
Investment per ton sugar of output (average season)	Rupees 2,305	Rupees 940
Investment per worker	Rupees 31,100	Rupees 3,530

Table 3.

capacity. From the table we can see that the investment per worker is nine times greater for the modern mill than for the low-cost mill. In table 4 a comparison of the cost of processing 100 quintals of cane is given. A quintal is about 50 kilograms.

Table 5 gives the output and employment resulting from the same investment, and shows clearly how the same investment in the low-cost unit gives rise to two-and-one-half times the output and eleven times the employment of one modern mill. The cost

Cost of processing 100 quintals of cane (based on 1971/72)

	Large-scale vacuum-pan factory	Small-scale technology
1. **Salaries and wages**	Rupees 164.35	151.00
2. **Fuel and power**	Rupees 57.70	66.50
3. **Stores and lubricant**	Rupees 103.23	62.80
4. **Repairs and renewals**	Rupees 48.77	12.00
5. **Depreciation**	Rupees 200.00	90.00
6. **Overheads**	39.23	10.00
7. **Taxes—a) excise duty**	123.50	59.70
b) purchase tax	50.00	50.00
8. **Cost of cane**	1,200.00	1,200.00
9. **Transport charges on cane**	47.50	—
10. **Capital cost 10 percent**	200.00	90.00
	2,234.27	1,792.00

Table 4.

Output and employment, same investment		
	(a) Modern mill	(b) Open-pan sulfitation plant
Initial investment (millions of Rupees)	2.8	2.8
Number of units	1	47
Investment per unit (millions of Rupees)	2.8	0.6
Total resulting output (tons of sugar)	12.150	30.280
Employment (permanent and seasonal)	900	9,937

Table 5.

of production per quintal of sugar is given in table 6 for both technologies. The percentage recovery is shown because of the difference in recovery rates of the sugar for the small-scale technology. From this table we can see that there is not much difference in selling price, and in this respect we have to keep in mind that the sugar produced by the large-scale process has to be sold in both near and distant markets, thus adding transport charges and local taxes. Furthermore, there is always a delay and time lag in the disposal of sugar from large-scale factories because it has to cover a much bigger region for marketing purposes. At the moment only 218 vacuum-pan mills supply sugar throughout India. These factors add five to seven percent to the costs, raising it to 250 rupees per quintal. No such expenses are incurred by mini-sugar mills because they are located near the farmers' fields and the cane is received in the farmers' bullock carts, while the sugar is sold on the local market.

Cost of production per quintal of sugar

1. **Large-scale vacuum-pan technology on the basis of 9.5 percent recovery**	Rupees 235.16
2. **Small-scale technology on the basis of 8 percent recovery**	Rupees 223.00
3. **Small-scale technology on the basis of 7.5 percent recovery**	Rupees 236.30

Table 6.

Some of the advantages which have directly occurred and have been listed by Garg are—

1. The units now crush nine percent of the total cane grown in India and produce three million tons of sugar, that is eight percent of the much-needed commodity of crystal sugar.

2. Capital investment of more than $48 million has been invested in rural areas.

3. Employment potential for 100,000 persons has been created in the agricultural sector in the slack seasons—the time when this labor used to migrate to cities to supplement its meager agricultural income.

4. Tax revenues have been added to the central and state governments amounting to $2 million.

5. Machine manufacturing industry has been set up of which the annual turnover is in the region of $10 million.

6. The requirements of iron and steel for the fabrication of machinery to produce the same quantities of crystal sugar are only about 60 percent that of large-scale industry. Thus 40 percent of raw material is saved.

7. More than 60 percent of the cane for the large-scale units is transported by trucks and railway wagons. Practically no such transport is utilized by the small units, since they are within easy reach of the cane-growing areas and carting is done by bullock power.

8. These units act as centers that can extend and make available mechanized facilities to the rural areas and can provide repair services for the new types of agricultural implements that are being introduced.

9. Large-scale technology sets up a trend for the movement of capital away from the rural areas. This creates a weak capital base in which improvement of agricultural technology finds it difficult to take root. Promotion of this unit has helped to build up capital resources and has many times served to provide agricultural inputs to the farmers.

10. The machinery design is of a nature that does not require any foreign components, thus saving foreign exchange.

11. This unit employs about three times as many laborers for the same capacity as does the large-scale technology.

12. The capital requirements to manufacture equivalent quantities of crystal sugar at this level are 40 percent of those of the vacuum-pan factories.

13. The price paid by these mills for the cane is similar to that paid by large-scale technology, and is at least 25 percent higher than can be obtained for other means of disposal of cane. At a rough estimate it can be said that an income of about $20 million has been added to the agricultural sector.

14. Several of the technological ideas of the small-scale technology have filtered through to the indigenous Gur and Khandsari industry; for example, the introduction of power crushers, the crystallization techniques, and improved boiling furnaces have been adapted with the result that quantity and efficiency of this old industry have also improved, resulting in economic benefits.

Other interesting work has been done by PRAD on whiteware pottery (for which a suitable kiln has been designed), spinning, and latrines for rural areas. Currently research is in hand on the feasibility of small-scale cement factories and improvement to the bio-gas digesting system, while in conjunction with other institutes, research is being carried out on the improvement of village spinning units and handloom weaving.

Brace Institute

The Brace Research Institute in Quebec, Canada, is located at McDonald College of McGill University. The particular reason for mentioning it is that it is a specific engineering institute that works in the field of appropriate technology, it has much direct experience, and it has evaluated much work that has been done in developing countries. The institute was set up by a bequest from Major James Brace, who was primarily interested in making desert or arid lands available and economically useful for agricultural purposes. It was his desire that the results of this research would be made freely available to all the people of the world. The institute was to concentrate on the problems of water and power scarcity affecting individual persons and small communities in arid developing areas.

A paper presented to the 1974 OECD conference of practitioners on low-cost technology stated that "it concentrates primarily on the technological aspects of these problems. It is fully recognized that a tool or system developed is only one facet of the problem. Full appreciation must be made of the cultural, social, and political context in which the equipment is to function, in order to establish its appropriateness to the community it will serve. The basic philosophy of the institute has been to develop saline water treatment. It has developed other energy-consuming equipment that maximizes local energy, material, and human resources so that the technology can find identity within the infrastructure of the local community. This policy was adopted in order to secure participation of the indigenous population in all phases of the construction and assembly of the equipment. This ensures continuity by developing their ability to handle its operation and maintenance."

In 1960/61 a research facility was built on the island of Barbados, West Indies, where abundant quantities of seawater, solar and wind energy combined to provide an excellent proving ground for equipment development. This overseas testing place, which lasted until 1967, and which is essentially being continued through expanded activities throughout the world, did more than just provide a convenient physical environment in which experiments could be undertaken; it provided an insight into the real needs of the rural population of the Third World. In surveying the needs of these rural areas it

Photo 11.4. Brace Research Institute airscrew windmill (Barbados).

Photo 11.5. Brace solar still for producing drinking water from saltwater in Haiti.

was evident that in maximizing utilization of local resources, considerable attention had to be paid to the development of alternative, indigenous energy resources such as solar and wind energy. As a result, development in the following areas has been studied—

1. Small-scale desalination equipment, solar desalination units, vapor compression, and reverse osmosis processes using wind power as the motive force

2. Direct application of solar energy for heating water, heating air, and drying crops

3. Solar-energy collection and storage

4. Solar-powered organic fluids, Rankine-cycle engines

5. Storage of thermal energy

6. The development of environmentally adapted greenhouses for arid areas to reduce heating requirements through a more efficient use of solar energy

7. Low-cost housing and the integration of solar and wind energy sources directly into the structure for the provision of services

8. Low-cost sanitary technology with a view to reducing water consumption

9. The development of a large windmill for water pumping, irrigation, or electrical generation purposes

10. The development of small-scale windmills using the Savonius rotor and sail wing principles.

173

Besides their research and development activities the institute has undertaken the field application of these technologies, for example, solar dryers to process corn for a feedmill in Barbados, a solar distillation plant in Haiti, and some solar cookers also in Haiti. It was in undertaking these applications in the real world that a need for a more comprehensive approach became evident. Enthusiasm accompanied by good engineering designs was not always sufficient. As described in the previously mentioned OECD paper, they reached a critical crossroad in moving from research and experimentation to implementation of technology in developing areas.

Through trial and error the institute has been able to set some basic objectives for their operations:

1. Wherever possible, local technologies should become involved in the development process in all its phases: research, development, and application. Hence, the institute has tried to help local technologists to appreciate the validity of studying the fundamental problems facing their own rural population. This is essential as they can communicate in the same language as the target communities and generally they understand the cultural and value limitations.

2. Local social workers are very important collaborators in getting the indigenous population to appreciate and accept technological innovation. For example, the installation of freshwater facilities decreases the infant mortality rate, leading to problems of birth control. Solution of this latter problem is often beyond the scope and capability of the well-meaning technologist.

3. Economists must also be brought in to provide a more comprehensive enumeration of the costs and benefits as they apply to given appropriate technologies in a local context. In view of past development experiences it is obvious that both the long- and short-run consequences of specific technologies need to be considered. These economists can hopefully specify more comprehensively the social welfare functions as they apply to given regions of a developing area.

Technology Consultancy Center in Ghana

The last organization that I want to mention is the Technology Consultancy Center of the University of Science and Technology, Kumasi, Ghana. I think it is especially worthy of study since I see it as a good example of what a university can, and should, do with respect to the development of technology for the greater part of society. The example set by the University of Science and Technology at Kumasi could profitably be followed by the universities of technology in the Netherlands and the rest of Europe. At the moment in the so-called "industrialized" world there is scarcely any effort in the universities to support the small- and medium-sized industries wherein the greater proportion of the working population earns its living. The TCC was started as a department through which the university could make available its expertise and resources to government departments, established industries, and individual entrepreneurs, but it has become more and more an agency for the stimulation of grass-roots development along the lines of the appropriate technology process.

A good example is the case of the entrepreneur who requested the assistance of the center in the development of a paper-glue from locally available material, such as cassava starch and the alkali from

Photo 11.6. Six-harness loom developed at the Technology Consultancy Center at the University of Science and Technology, Kumasi, Ghana.

the skin of the plantain. The center provided the technical know-how as well as building the production plant. It also gave advice and help to the entrepreneur in obtaining a financial loan. The result is that within a year of the start of this project the entrepreneur is supplying the best part of the country's requirements of paper-glue, thereby saving a substantial amount of foreign exchange. Additionally, this newly developed industry is providing employment to about 20 rural dwellers who otherwise would be unemployed. A planned expansion is expected to make Ghana self-sufficient in paper-glue and thus bring to an end importation of it into

the country; indeed, prospects for exporting it are very bright. The success of this small industry is likely to lead to the development and adoption of other technologies. For example, the glue containers that are used at the moment are imported plastic cans which often run short on the local market. The TCC is investigating the establishment of another industry to supply suitable containers made from a material of which there is a dependable local supply.

Although this is a good example of harnessing a university's technical expertise, it is noteworthy that after the development of the glue formula, most of

175

Photo 11.7. Soap plant developed and built by the Technology Consultancy Center. It is capable of producing 1,000 2½-pound soap bars per day and is operated by a cooperative.

the remaining work of establishing production was carried out by the entrepreneur, who kept in close contact with the center. From their experience however, they have learned that not everybody keeps in touch and provides them with feedback, and that their technical advice is by no means necessarily acted upon. It appears that entrepreneurs can be unable or unwilling to translate the oral and written technical advice into concrete results; in this sense at least, the paper-glue project may well have been exceptional. It became obvious that in some cases there was a need for the center to become actively engaged, not only in the drawing up of plans and in feasibility studies, but also in undertaking pilot projects. This active role has two objectives. Firstly, it can demonstrate the viability of such projects in technical, economic, and other terms. Secondly, it can provide facilities for involving or training interested parties.

Examples of the industries that have been initiated by the center acting in this way include soap and caustic soda, steel bolts, school equipment, cloth weaving looms, and others. The following list further illustrates the range of technical problems brought to the center by small industries:

1. Search for a substitute for linseed oil in putty
2. Improvements in the local manufacture of gunpowder
3. Testing of powdered and liquid soaps
4. Testing and manufacture of a bleaching fluid
5. Production of jams and fruit preserves
6. Installation of a cold store
7. Wig manufacture from vegetable fiber
8. Refining honey
9. Search for a substitute for cork for gaskets
10. Inspection of a lathe damaged in transit
11. Electrification of a flour mill in a rural area
12. Production of egg boxes
13. Analyses of a brewery's spent grain
14. Softening of lead for making fishing weights.

Conclusion

I have attempted to focus our minds on the way that technology is employed in our own, industrialized, societies. Clearly a vigorous analysis is right outside the scope of a lecture like this, and all I could do was to show that a very high human price is being paid to sustain our life-style. However inadequate such a view of "sophisticated" society may be, it is important to bear in mind that this way of life is very precarious. Quite apart from the stresses that resource shortages are subjecting it to, there is a ground swell of dissatisfaction and alienation from people who feel themselves trapped by the inherent competitiveness of the system. This ground swell is, I judge, gathering momentum daily.

I said that I could only ask questions about our society and our hopes for it. I can provide no answers. But I think that if we look to some of the organizations in the underdeveloped world that I mentioned earlier, we may begin to see some answers for our own society. The criteria for a socially appropriate technology surely have universal application. And the philosophies and methodologies that have been evolved to combat material poverty in one setting are certainly capable of adaptation and improvement to combat societal poverty in another.

Chapter Twelve

The Transfer of Knowledge and the Adoption of Technologies

by Harry Dickinson

"A student who can weave his technology into the fabric of society can claim to have had a liberal education. A student who cannot weave his technology into the fabric of society cannot even claim to be a good technologist."

These are the words of a British colleague of mine, Sir Eric Ashby, and I think they get very quickly to the heart of the problem. If we as technologists and engineers lose sight of the human end that is the purpose of our work, we would be reverting to something like a medieval theology rather than practicing technology.

I shall start by looking at the relationship between rich and poor in a rather general way—with what I hope may be called an inquiring mind, for an inquiring mind is the best, and probably the only, asset that one can take to the problems of development. Much of this view will be focused on technical education and is seen from the perspective of an electrical engineer. I shall try to describe a little of life in the Chinese countryside and the communes which are, I think, the most exciting experiments in social organization which the world has seen for a long time. And finally I shall try to make some reflections on developments in other parts of the world in light of the Chinese experience.

Price of Education

The developing countries, even though they are much poorer than we and cannot support our institutions, want to obtain for themselves the apparent results of our institutions. Consequently they build up universities, and other technological institutions, which are copies principally of European, sometimes American, ones. They finish up with an institution which is well geared to educate their people for employment in Europe and America, and is absolutely irrelevant to the countries they are actually working and living in. Then they come to join our professions which we don't yet criticize much, although I think that a harsher look at the relative remuneration, the relative importance, of professionals is about to take place in Britain.

Professions

Let us look closely at the role of the professions and what they do. To take just one extreme example, consider the dental profession in Britain, which deals with teeth that are faulty, that are decayed, that need replacement. It never deals with the more fundamental aspects concerning teeth. It is a "theology" concerned with bad teeth; it has nothing to do with healthy teeth. It seems that we are incapable of using the products of the sugarcane rationally for producing a lot of useful materials; we arrange things so that it rots the teeth of our children. Then we "invent" a profession, which we pay highly, to put them right again. Now, this is an extreme case, but if you look at our medicine, or at much of our engineering, we have invented the problems so that we can solve them with a capital-intensive solution. They have little relevance to reality. Now, subject to our influence and our own lack of self-criticism, the poor countries of the world are building up institutions on this sort of pattern to turn out their own professionals and other highly educated people. I find this worrisome, because even we are finding it increasingly difficult to support this unnecessary superstructure.

As an inhabitant of one of the poorer countries in Britain (Scotland!) I am sure we are going to learn how to deal with the professions before you; you will learn the lesson from us. We find, for instance, that people's motivation in Britain is no longer for well-paid jobs, nor do they want importance. Students don't want to join the establishment in the regular sense. Over the last few years they have revolted against this violently—not quite as violently as in other countries, but violent by British standards. We have now passed that stage and students are seeking a new motivation which none of the professions can give them because they can't understand it. Students doubt if they can be of much use to the world, but they would like to earn their living in a way that doesn't harm anybody. Young people in Britain and elsewhere are looking at the problems of the world in quite a new way. It is now perfectly logical and reasonable for anybody who is young to say, "Right, if we are short of fuel, why don't we just use less? If we can plan for all sorts of other things,

why can't we plan for the use of less fuel? If the conclusion is that we must promote public transport and have fewer cars, then what are the techniques of achieving this situation?" Arguments about the advisability of this sort of conclusion seem to have disappeared; it is becoming a necessity. I think our students are so far ahead of us that they are now teaching us in our classes. They have evolved a world view, which I find coincides with the one I got by working in developing countries, and not by seeing the reflection of these countries in the British scene.

Get Off Their Backs

Assuming that we see we must change our values, we must also realize that for a long time we have looked upon the fact that resources are badly distributed and badly used in the industrialized world as an "act of God." There is nothing we can do to benefit the poorer countries of the world. We cover our consciences by saying that the reason the poor countries can't get anywhere is not that we use all the resources; it is that they have too many people. And this, we find, is a self-satisfying argument.

The conclusion to draw from this position is that the easiest way to help the developing countries is to get off their backs—not to use their resources, or, when this is unavoidable, to use them more intelligently, to buy processed goods from them so that they, rather than we, benefit from the added value. We should make sure that we get poorer so that, with the limited wealth of the world, they can get richer. This, you will agree, would not make me popular with any politicians. But it is the logic of the situation. If we cannot face these problems among ourselves, we have no hope of doing anything for the developing world. We can no longer think of charity, nor can we believe that we are doing anything to help anybody other than to salve our own consciences. So my first message is, I think, fairly negative.

Now presuming that you survive the shock of this conclusion—and the human race is very resilient and has so far been able to beat all the systems that have been devised to ensnare it—and assuming that we are politically able to put our own house in order, is there really a useful role for us to play? Is it possible that our abundance of professional, technical, and administrative skills and talents can be useful to societies that haven't yet been able to establish education and training to meet their own needs and aspirations?

Prestige

To illustrate my views I'll look a little more closely at the field of technical education, but I believe the argument can be generalized fairly well into other areas. The desire for technical universities and colleges arises from both the obvious utilitarian needs for technologists and technicians and from the prestigious nature of such institutions. A developing country is apparently driven to having a spacious campus, a nuclear reactor, and an international jet airline. They feel undressed until they have such organizations and are without any apparent concern that the benefits are in no way related to the costs of such things. We could make exactly the same analysis of a lot of hardware, from Cleopatra's needle through to the Concorde. Nobody in his right mind would ever build a Concorde if it were not for prestige purposes, because who would proudly say, "I have invented a method of getting across the Atlantic which needs two-and-a-half times as much fuel per person as has ever been used before," in the face of a world of diminishing fuel resources? Yet the British and French governments are committed to

their own financial disaster; I think that long after we have lost a few universities in Britain we will still have Concordes.

On the other hand, vocational institutions suffer from a lack of prestige, but in general they can be more readily planned to meet ascertained needs. They suffer, however, from the problem of retaining staff, because even though a lower-order institution may be much more useful and appropriate, its staff can see the chance of promotion to one of the other institutions. I think there is no country in the world which has solved this problem, with the possible exception of China and I don't think it has been fully successful. We have to accept this situation as a continuing and inevitable feature of all but the most pragmatic states. And by pragmatic states I would include China, although both Algeria and Cuba would fit the pattern to some extent, but the state monopoly (capitalism?) of our East European colleagues strikes me as displaying exactly the same features in its institutions and capacity to transfer useful information as does private monopoly capitalism.

We find that when we want to be of use to developing countries, the ones that are pragmatically concerned with developing their own resources say, "Thank you very much, we don't really need your help; we will stand on our own feet and do it by ourselves." Once they have established their competence, as China has done, they say, "We know we can do it, but now maybe we can reach an arrangement with you to make our lives easier and bring us closer together in the world community." But they have to prove that they are independent to start with.

Honest Intentions

But what can we do for the less determined independent states? Assuming that we accept all the limitations of our own societies, what possible assistance can we provide to institutions of technical and higher education to make them useful in the national sense? For you must remember that the American university criteria—money donated by big business and the number of Ph.D.'s graduated—hold all too often in the rest of the world. If we wish to persuade those in power in developing countries that we are involved for the benefit of their people, and not to serve undisclosed motives of our own, we have to act on our own beliefs. I have to say this, because most of the poor and the apparently unsophisticated of the world know that Europeans are rich, clever, and cunning. And that we have managed to stay that way for a long time at the expense of the poor. The Mexican peasant or the Andean Indian knows that this is true, and his analysis strikes me as much more fundamental than that of the Mexican or Columbian intellectuals with whom I have worked.

Or again, the people of West Africa have seen their social organization disrupted in the early days of colonialism by an irresistably powerful mixture of the repeating rifle, Christianity, and gin. And the combination of these three was so basically socially disruptive that it took a very long time to get over the combined effect. The exploitive aspects of colonialism in West Africa were as nothing compared to the social disruption of those days. No sooner had they got over this than we invented DDT, and gave them a population rise with which their culture could not cope. Finally, as they are coming to terms with population, we have invented higher education, and this again is maintaining the social disruption.

So they have a lot of reason to be suspicious when we say, "I have a nice cheap way of doing something; we wouldn't touch it ourselves at home but it might be the right thing for you." We are the ones who have to demonstrate our bonafides; we should not be offended when they don't believe a word we tell them. Indeed, I would say that if we are working in a community that doesn't believe what we tell them, then they are pretty bright and will be excellent people to work with. We must try to show that we are doing something that matters to them and that we really have something to offer.

Three Principles

If we can get over this, there are three principles we have to accept, which are hard to come to terms with, since it is our own culture that is threatened by them.

Firstly, while acknowledging the universality of the laws of science, we must extend consideration of the application of our scientific knowledge to a wider range of situations than is encompassed by the experience limited to our own society. This principle is difficult to accept because, having high capital intensity and being somehow economically in the lead, we presume that we must be intellectually superior to those who aren't. We have to show that we too are ignorant and wish to learn, if we want to operate in the development situation. When we look at societies outside our own we say that other people have no technology, or that it is a crude, or primitive, technology or that a technology that did not solve our problems will maybe solve theirs. This is the usual sophistication of our arguments and I can produce endless United Nations documents showing essentially these views; it is something we ourselves have to change.

Secondly, we have to specify in some form that the problems of poor countries are as important and as intellectually taxing as the problems of the developed world. Again we have to do this by ourselves. It is very important to recognize that these so-called poor technologies have enabled societies to survive until modern times for one, two, three, four, or five millenia, and they will continue to do so without the need for outside inputs. We have had our own technological base for 100 years, and know it can't sustain as unmodified for another 100 years. So someone from an undeveloped society would have cause if he wished to be arrogant in this situation. We have to come to terms with that.

Thirdly, the solution of technical and economic problems in underdeveloped countries should be made in the context of the optimum use of material and human resources in such countries, and not by the implantation of processes optimized to the resource pattern of a different society.

Having gotten this far, we find that we are left with a very broad knowledge of the laws of science, a few tricks of the trade which are of fairly universal application, and a lot of ignorance. We could overcome the ignorance obstacle by saying that we can't be ignorant—after all, we've been to the university and have degrees and certificates saying that we aren't. But looking at it more closely we have to admit that we are ignorant. Take, for example, the scientific and technical research and industrial development carried out in towns like Eindhoven throughout the developed world. It has been estimated that certainly 95 percent, more likely 98 percent, of all scientific and technical research and development is concerned with the problems of the developed world and not with the poor world; this leaves something like 2 percent concerned with the poor world, and of that only a part is carried out in the countries concerned. So, at a rough estimate only 1½ percent of the world's research is done in that two-thirds of the world where the population has real problems. The other 90+ percent is done as a cultural exercise by people like us who in-

vent problems so that we can prove we are intellectual giants in tackling them.

Generate New Technologies

To meet the needs of these people I think there must be different technologies, because their problems are different from ours. They also want to survive and to not find survival difficult as we shall. Now, I cannot define a socially appropriate technology. All I could do is write down a lot of characteristics about meeting technical, social, and economic needs in terms of indigenous resources, indigenous qualities, and indigenous social structures, rather than in terms of any reflection of our own social arrangements.

Certainly we do not want to be acting as experts, or advisers, or the people who manage projects. Rather, we should be working to extend the freedom of technological choice. Most people in poor countries know their own practices very well, and they know ours. But they are unaware of criteria and have little information on any other alternative. So if we want to be involved honestly, we have to move in this no-man's-land, looking for alternative techniques which fit the social and cultural background we're working in and which meet the factors of production in the widest economic sense.

Modify

There are broadly three ways to generate such appropriate technologies. The first is by modifying the existing practice already used in these countries. If we find they have design limitations because of lack of knowledge of materials, if the processing techniques, tools, or traditional ways of doing things can be improved by small modifications which we can understand because of our superior knowledge of scientific processes, then we can produce fairly quick improvements. They will not be spectacular; people will not grow rich overnight on the basis of technology of this sort.

Maybe people should never want to grow rich overnight; I was looking recently into the social and political history of Britain. The whole idea of growth is such a curious notion, rather like a disease on a stable structure that only appears now and then. We had a period of about 400 years when the g.n.p. if anything declined slightly, rather than rose. It was only at the end of that period that we had the economic changes which led to changes of economic standards, changes of wealth, and changes in people's expectations. I think that growth will be fairly ephemeral in the history of the world. Although scientific understanding may enable a better quality of life for more people, continuous or sustained growth, or very high energy consumptions, it is simply not possible in the long term. Many centuries before our economists talked about growth, we had in Britain the Black Death plague, which halved the working population and so doubled the g.n.p. per capita. This method is still available to us today but is not looked upon as socially acceptable—there are social restraints even in the fastest growing economies.

Long stretches of history can also be seen in this fashion. There are areas of China where the standard of living and the population have been stable for about five millenia. This is based on very small changes of technique but with a grasp of certain fundamental aspects of science, such as coping with plant diseases, with changing varieties of plants, with changes of climate over time. There was quite a lot of indigenous technology, but because it was not recorded and because there are not courses on the technologies of the Imperial dynasties of over two centuries ago, we

tend to believe innovation never happened. But their technologies are as real as ours, and there is no reason why they cannot go on for another five millenia—ours can last another 100 years if we're lucky.

So, if we generate an appropriate technology by modifying existing practices, we must do it in such a way that the local people make the decisions, that we really do give options, and that the indigenous structures of social, political, and economic control are real and can prevent too much unregulated change stemming from any innovation. Small increases of productivity or changes in minor parts of the economic activity are quite dangerous. One of the panaceas we have suggested to the Third World in general is mechanization. "Mechanization is good for you," was roughly the message of the first development decade. As a result more people are starving, and there are many rusting tractors which fortunately broke down before they had completely destroyed the community that received them.

We have seen this same immense optimism in the Green Revolution which produced more rice in India at the same time as it produced more poverty and made the rich a lot richer. The lasting effect has been that the concentration of land, wealth, and power has increased in India, and so has the number of people who cannot afford rice. This has nothing to do with our usual reservations—that the technical innovation was not a good one, that it was not sensible, that it wasn't ecologically sound—but it was quite unsuited to the social and political structure into which it was introduced. When we introduce one-shot pieces of magic we must expect to find social ills. Even the unsophisticated Andean peasants know a lot more about the meaning of failure to survive than we do. In the last resort they are the experts of their own social and cultural conditions, and we ignore such wisdom at our peril. So, any alternatives we offer must have inbuilt safeguards for the social structure.

Modernize

The second way we can help is to go back to our own industrial history and reexamine what would now be called labor-intensive, ecologically sound, simpler, small-scale processes which were developed in our own technical revolutions. We could possibly refurbish them and find that their factor proportions were more realistic for African or Latin American peasant societies than the other alternatives we can offer at the moment.

This approach is an interesting source of possible alternatives, and we will find that these technologies were abandoned not because they were uneconomic, not because they were socially disruptive, not because they were ecologically unsound, but because we had a sudden belief in the economies of scale (if things were bigger somehow they must be better). The 1851 exhibition in Britain was the great monument to "progressive thought"—to change things to be bigger and better and more energy-intense as a virtue, quite irrespective of the actual production demands of the situation. The optimism of the European bourgeoisie, in particular the British Victorians, followed quickly by the French, very much influenced the invention of economic theories to account for what had actually happened. You will gather that I am a firm disbeliever in the supposed science of economics. I think it is always invented in retrospect and is then used to predict from inadequate data, explaining its almost universal lack of success.

Invent

The third way of proceeding is to invent a new technology, or to change the scale of a modern technology in some way we have perhaps never before experienced. This means trying to use our own ingenuity and our modern knowledge so that people can utilize their own raw materials without generating mass unemployment and without producing social disruption. I recognize that any society will change its social structures in time and I am not advocating that we try to invent things that turn people into museums. But we should invent things that give people control over their own capacity to change, and the way they do it. There is no universal virtue in reverting to the primitive savage, having no sanitation, and dying at the age of 33. I'm only suggesting that we pass the freedom of technological choice back to the people concerned.

Perhaps I can show how this viewpoint is reflected in the educational field, since these considerations must have a bearing on education at all levels. If in West Africa I want to describe the elastic properties of materials, or the concepts of stress, strain, elongation, and modulus of elasticity, I can explain them as general concepts of universal application, which people with a little familiarity of the analytical way of looking at things can understand wherever they come from. But for students who are not totally familiar with these concepts I would have to draw on materials they know.

In developed countries I would naturally talk about steel, aluminum alloys, plastics, and I might include wood somewhere at the end. In West Africa I found that I had to start off with the elastic properties of bamboo—a very familiar elastic material. But, even though I could find in all the textbooks the elastic properties of a rod of uranium, should I happen to own one, I could find nothing however crude, of the elastic properties of bamboo, in terms of its use as a structural material. And after all, half the buildings in the world are probably built with bamboo; it has caught on better than steel and glass with most people. When I looked at domestic fuels I found that wood, charcoal, or rice straw are the most widely used, but again there are no calorific values in my textbooks. Coal is virtually unknown to most people in tropical countries, yet most of the textbooks are full of it.

On another occasion I was trying to teach in West Africa about overhead transmission lines following a syllabus that had been bequeathed to them by London University. I arrived to find all the students busily trying to calculate the effect of ice-loading on a transmission line, and discovered that the only ice that they had ever seen in their lives was what came out of a refrigerator. The concept of having a tube of ice on a transmission line was very difficult for them to imagine. They wondered how we got it there! These were students who were going to get British-style honors degrees. As I have said, we were going to give them certificates saying, "You are useless to your own country, but we'll find you a job in ours." This attitude, I think, is very dangerous.

Technology Consultancy Center

I want now to turn to ways in which existing institutions, particularly educational ones, can in practice adapt themselves to embrace some of these considerations. In the University of Science and Technology at Kumasi, Ghana, I was involved in this type of evaluation and we began by asking, "What are the resources of the university in terms of the needs of indigenous forms of production—village industries, small industries in the towns, state corporations, or even the big expatriate firms? What do they do and why? Can we adjust our training to meet these needs? In this training

can we, at the same time, try and solve some of the problems for the poor of the country?" The reaction seemed fairly hopeful, and we persuaded somebody to put some money into an organization which we called the "Technology Consultancy Center."

The center operates in villages, in small firms and big corporations; it also charges fees which we think are socially realistic fees for the enterprise. Very roughly, any big firm, expatriate company, or government body is charged the same consultancy fee it would pay a European firm, on the grounds that they'd have to get them in anyway and it would probably cost them more for all the incidental expenses. The small firm or businessman in the money economy, is asked to pay a small percentage of the increment of his profits for the next few years, or a lump sum; the choice is his. And the villagers, who are largely outside the money economy, are asked to pay something. So two yams might be a realistic social fee for actually designing and building a pump which cost a lot more in man-hours. This system is fairly acceptable and people now travel perhaps 50 or 80 miles on bicycles from the middle of the bush with a bit of broken equipment saying, "Can you mend this for us?" This I think is a reasonable show of success.

The center has also managed to start one or two small industries. For example, traditional soapmaking has been revived in the villages to compete with Lever Brothers and others. It has been possible to make Ghana self-supporting in glue and thereby to save it having to import glue; not that I think it is an entirely good idea, because the main users of glue are bureaucrats of which there are already too many. Some of the center's work has had some influence on the university curriculum, and some association of the students with a number of these little

185

projects has been possible. It is hoped there will be a continuing spin-off from such activities into teaching.

School Equipment

One of the activities might give ideas to any electrical engineer who is looking for things to do. One of the local teachers pointed out that the biggest trouble with teaching any form of science there was the exorbitant price and the ridiculous design of most teaching apparatus for schools. So a unit has been set up which has given itself the splendid title of the "Schools Science Equipment Imports Substitution Organization." The teachers used local woods to make things like meter sticks, retort stands, and worked their way up to producing resistance boxes, potentionmeters, and various sliding wire devices. In order to make coils on a former they had to find the cheapest piece of wood that was "tropicalized" and reasonably dimensionally stable. The answer was Chinese pencils; they sell at about a cent each and have flooded West Africa. So they were all bought up to be used as formers for the electrical devices.

The import substitution group has gone on to make simple amplifiers for teaching purposes and has worked out a number of simple instrumentation rigs. It cooperates with other active people, particularly pharmacologists who are looking closely at indigenous plants, and have tried to produce simple ovens and dryers, pH meters, incubators, constant temperature baths—a whole range of apparatus which can be built using local material, minimizing the import components as far as possible.

The TCC has started a manufacturing unit on the university campus. The use of steel coach bolts for building vehicles in Ghana is relatively limited and the center devised a labor-intensive, rather crazy, method of making them, which is cheaper than importation and can be done with local skills. It is not an optimized method in any way, but it turns out bolts having two-thirds the strength of the imported ones—and the strengths required are very much less than that. Production was started on the university campus, and two technicians have taken it over and run a small business employing about 17 men.

I find it very cheering that if you see a Masai tribesman in East Africa these days, you will know that the tip of his spear was made from the leafspring of a Ford car. Word has gotten round the continent that springs have the best steel for weapons and there seems to be a premium on Ford springs. Indigenous blacksmiths found this out and modified their forging and treating methods, using very crude traditional designs, to use a material of which they had no previous knowledge. They handle it fairly well; obviously we could improve the techniques if the use was serious enough to need the full properties of the material. But as it stands at the moment the fact they are doing it at all is very encouraging.

Tanzania

Before moving on to talk about China I just want to mention some aspects of Chinese aid schemes in Tanzania. The Friendship Cotton Mill is well known but nevertheless is a really good example worth studying. I was told that the machinery presently used in the Friendship Mill certainly came from China, but its history goes back a lot further than that. This equipment was being run during the war in Germany; it was taken as reparations to the Soviet Union and was refurbished and passed on to the Chinese, who used it for many years. They, in turn refurbished and passed it to the Tanzanians to start their industry, before they go on to slightly improved Chinese models; and it is already earmarked to go to the Somali Republic to start a cotton industry there. The mill dates back at least 30 years and yet with fairly clever use and concern for repairs and maintenance, particularly in China, it still performs well.

Perhaps this story exemplifies the strengths of appropriate technology. And even if the mill will not be the ultimate of their cotton industry—they will choose for themselves from the range of equipment available—in terms of starting a new industry, the older technologies are not a bad introduction that can give people some freedom of movement.

I didn't in fact get into this factory but next door was an agricultural tool factory which I visited with a dozen students from the university. I was one of the few pale-faces in the party, and much to the amazement of the African staff, the Chinese all rushed over to shake hands with the strange visitors. After we'd all stopped laughing we discovered that the Chinese had imported the equipment and had actually trained Africans to use it without any of the Africans knowing a word of Chinese, nor any of the Chinese knowing a word of Swahili or English. It had all been done by example, including the spectroscopic analysis of steels. It had been demonstrated entirely by sign language and shaking hands and patting each other on the back, or pulling faces when it didn't work. It had been slow but there was a complete rapport, and the plant worked very well. And it was the first African factory I had been to where, when the plant failed or broke down, the ordinary operatives downed tools, picked up the repair tools, and started repairing entirely on their own initiative and entirely under their own control.

They only had two complaints in this factory; one was that the Chinese were pulling out after five years, and the African workers had never discovered which one was really the boss, because they all worked together. But within two days of the African boss being appointed, he had a large car, a fancy shirt, and an air-conditioned office. I am sure you know where he got those habits from; they are not indigenous to Africa. It is obvious that he had copied what goes on in European-run factories in East Africa and thought his status demanded these things. We had obviously educated him into these cultural values which we think are laughable and which look completely ridiculous there. This poor man had been conditioned by us, in the guise of education, to react in this fashion.

The second complaint was that the Chinese scrupulously collected all the scrap from this factory and stacked it in various grades. Once every two or three months it was shipped back to China, when there was an empty boat. At the end of the year they said, "You see all that scrap, in fact it is an input for another process and is valuable. We have taken it back to China to use, but it is now up to you to sort it and invent your own sideline industry to utilize it." Now, unfortunately, outside this factory is a splendid pyramid of mixed scrap, rotting gently in the African rain. Nobody there was able to take that sort of decision, to do the experimentation or to attempt to utilize it. This contrast between the confidence of the Chinese and the lack of confidence of the inhabitants of the former colonies is, I believe, one of the most marked features you can see in almost any part of the world.

China

I think perhaps that many of the points I have been trying to make will become clearer as I describe a few examples of the way that technology is valued and utilized in China. In no way does this pretend to be an account of technology policy or practice in China, still less an evaluation of the Chinese "system," but I hope it will show technology in a different light—that it is perfectly possible for technology to harmonize with the social, cultural, and economic situation of peasant communities.

China's urban centers appear to suffer the same sort of problems as cities everywhere. The Chinese admit that it is deplorable but say they can do nothing with it except wear it out and, when it is rebuilt, try to make it a more rational and sensible place to live and work in. They are not rich enough to scrap it and start again with "ecological" industries, but they are aware of the problem and hope to learn from past mistakes.

By contrast the countryside appears much more like one expects China to be. The only motive power used, apart from man, is the water buffalo. Everybody involved in agriculture works in a highly labor-intensive way, as you can see from the number of people in the fields. This sort of scene is exactly what one would expect to see, until one realizes that everywhere one also sees concrete posts for electricity distribution.

There are three inputs that the Chinese government sees that it must give to the communes to make them autonomous and, as one example of this autonomy, to make them free to select their own technologies. The first is national and regional control over flooding and irrigation. A controlled and ensured water supply is considered vital. The second is the provision of relatively modest amounts of fertilizer. Some chemical fertilizers are needed in China to top up total amounts available or meet deficiencies in areas where most that is used is organic. The third input is the provision of electricity. This is seen as the key that can give the communes flexibility in what they want to do. Electricity not only makes

life easier, but it is a simple means of allowing diversity in a whole range of mechanical and other processes that the communes can undertake.

Electricity

The Chinese manufacture a wide range of electrical apparatus, from small motors that are made on the communes right through to 300-megawatt hydrogen-cooled generators which you can buy off the shelf in Shanghai. We visited one commune not far from where large generating sets were being made. Some of the people in the commune became inspired by this neighboring industry to do something electrical. After much argument and persuasion that they were capable of learning the skills and could actually manufacture equipment, they were allowed to bring in the raw materials and equipment they needed. They have a diverse range of second, third, and fourth-hand machines that have been sold by various factories to those communes that can use them productively. The communes are now building four-kilowatt motors, of which they produce something like 8,000 a year. This is done entirely with peasant labor—none of them actually had a formal training in mechanical or electrical engineering. They simply took their enthusiasm and learned enough to do the job.

Although they have such a wide range of electrical industry, the Chinese see nothing strange in having old and new, large-scale and small-scale technologies which are integrated into the whole economy. Elsewhere I have described the situation as contrast, not conflict, between old and new. This sums up very well the attitude towards innovation and pragmatism. Such an attitude can be seen in many contexts. For example, a buffalo can plow much less land in the course of a day than can a tractor. But in a commune the man who drives the buffalo would be paid slightly more than the tractor driver, because the man working with the tractor has an easier time physically. The remuneration has nothing to do with production; it is based on the effort believed to have been expended for the common good.

Electricity has been distributed almost in the form of mania throughout China. The bulk of the generation is done by local hydrostations of about 2 megawatts capacity. Communes run their own generating stations—anything from 50 kilowatts up to 2 megawatts. This is the basis of the country's electrification, apart from the coastal cities which use coal, a small number of stations fired with oil, of which China has found a fair amount, and some

large hydroelectric plants which are appearing on the great rivers.

A grid system is spreading fairly rapidly all over the country and transmission lines for the lower voltages are made and erected by the local communes. Sometimes they use concrete posts, sometimes any handy bits of wood. Quite often the porcelain insulators are made locally too. High-voltage insulators are rather long-run insulators which they seem to be able to fire by relatively primitive methods better than I have seen done anywhere else in the world. You even find such sophistication as surge diverters put on the ends of the lines where they come into the transformers.

One aspect of their transmission lines which is instructive is that they use guy

Photo 12.1. Children in Shanghai learning basic electrical work.

189

wires to support the posts much more than we do. The reason that we don't use guys is that we have private ownership of land, and you have to pay more way leaves when taking a line through a dense agricultural area if you use guyed posts than you do if you use towers. So we use about three times as much steel as the Chinese do, just because we have invented a social situation called "way leaves." This is another example of the inappropriateness of much of our technology.

Village Electrification

One sees many houses that have been or are being connected to the electrical system. Individually the first priority for using the supply is, naturally enough, for lighting; perhaps an electric fan might be bought later when it can be afforded. Communally, however, the priorities are more utilitarian—irrigation pumps as the first thing, followed by grain huskers, maize grinders, and so on. One village we visited had just gotten electricity and they had literally taken as their priorities: first came the pump, second the grain miller, and third the television set.

In China it is quite appropriate for traditional and modern technologies, such as the dragon's spine pump and the electric pump, to exist side by side.

The dragon's spine pump lifts water through a head of about one meter, through the action of flat paddles pushing water up a trough as the handles are turned. Two men can operate such a pump for about 20 minutes at a time, then another pair takes over. A team of our men would work in this fashion for a six-hour working day. Between them they might be generating the equivalent of as much as 200 watts over this period. But when a commune gets electricity and can afford a pump, which is fairly cheap in China, it might buy an electric pump. Not only is this more efficient but it is powered by a four-kilowatt motor. In other words, it can do the work of 20 dragon's spine pumps and 80 men.

Where a village cannot afford or does not give priority to electric pumps, you will still see the landscape full of people. But the scene where electric pumps are in use is quite different. They are easily installed—bamboo poles are erected to carry the three phases in. Areas like this will have a skyline with nobody at all.

The contrast in the Chinese scene, caused by such minor innovations as electric pumps, is really quite remarkable; the real power of mechanization or industrialization in terms of what it can do for good or for social evil becomes apparent very rapidly. This is exactly the sort of thing that must have happened in Europe when we started to mechanize. Although we have seen the disasters that have occurred elsewhere when mechanization caused mass unemployment, we were lucky that a reasonable balance was kept by the demands of industry absorbing our population. All Europe's problems whilst it was industrializing were as nothing compared with the problems of countries that went through this process more recently, or that are still going through it. This is because our big population growth didn't take place until our industrial expansion was substantially underway, by which time it was capable of mopping up surplus labor as it went along. We are, of course, now getting into second-order difficulties because we have too many people for the types of industry we run.

As I have explained, China is feeling the effects of mechanization. If you ask the people in a village what happens when they get a pump, they say that they all work less hard. But that is only their immediate reaction to the innovation; after a while they say that they start up new "sideline" industries. These may give

Photo 12.2. Snow removal in Peking.

them the added value from processing their own raw materials. They may produce goods that the community itself needs, or they may produce goods for sale and thus enable the community to purchase products it had not used before. Whatever the objective of the new activities, there is no idea that anything to do with increased productivity could lead to any sort of unemployment. Such a result is inconceivable.

Local Industries

One "sideline" industry, which has been established to absorb unemployment resulting from mechanization, is the shoe factory. Women who own their own sewing machines come together for four hours a day to run a factory for making shoes. Perhaps another group might come together for embroidery. This is a collective activity so that young women with children can make a contribution towards village productivity when they cannot participate in the normal full-time activities.

Many of these little factories have been established. Moreover, these are social organizations; the people involved are often related to one another. And in these industries, there is no sign of any management. People simply work together and chat together, trying to be productive at the same time.

Another aspect of Chinese industry involves the sensible use of resources. The threshing of rice provides a good example of this practice. Customarily, rice is threshed by a stone in the shape of a truncated cone which rolls around a circular path pulled by an ox or pushed by half-a-dozen men. The winnowing is then done by hand. But the straw that is leftover is

191

Photo 12.3. Bicycle parking lot in Canton.

Photo 12.4. Mill in southern China.

collected and is either used as a domestic fuel or it is taken to the brickworks and used for firing bricks.

As all of us trained Western technologists know, rice is one of those plants that collects silica from the soil. And when rice straw is burnt, its ash is very rich in silica. The Chinese peasant knows this too, and knows that it is therefore a pozzolanic material. So when rice straw has been used for firing bricks, its ash is added to cement to produce a better quality or to extend the cement in some way.

Ferrocement

A lot of the ash is used in ferrocement, which the Chinese utilize more than most people in the world for a lot of different purposes. Ferrocement structures are made from a wire-mesh shape that has a very fine cement mortar mix spread over it; the material has very elastic properties. In the West we know it as a material that rather way-out architects, builders, and designers are proud of. It is mostly used for prestige buildings of which the Sydney Opera House is a good example.

In the Wusih area of China it was decided many years ago that reinforced concrete was not of very great interest to them, so the principal building material continued to be the traditional brick. But this is one of China's canal areas and when it was realized that timber for making boats was becoming very scarce, they started to search for alternatives. The factory that we visited had been a building materials factory. Even though none of them had built a boat before, they were asked to have a try at

Photo 12.5. Bean curd shop—Shanghai army base.

Photo 12.6. Making ferrocement sampans at the Horse Bridge People's Commune, China—stitching wire mesh to the underlying frames.

building one from cement, since they had the most skills in using this material. In fact they built two. First of all, since they knew how to build sheds, they made a shed out of cement, tipped it upside down and launched it to see how it floated. Then, for comparison, they got a wooden boat and cast cement round it. On the basis of the performance of these two strange objects, they modified their design. In this province now they are turning out about one million tons of ferrocement boats a year. The boats are about 25 percent heavier than wooden ones, but this is largely because the Chinese safety standards for small boats are somewhat higher than those of Britain. Boats designed for use on their inland waterways would be considered by us quite suitable to cross the ocean between Britain and Holland.

A 50-ton fishing boat, for example, is extremely sturdy and seaworthy. Its superstructure is made from wood but everything else, including the crossbeams and bulkheads, is made from ferrocement. This size of vessel is now commonly used on China's lakes and could well serve on many rivers and seas throughout the world.

Since I started with a quotation, I shall finish with one. It is from Chairman Mao. "Now there are two different attitudes in learning from others. One is the dogmatic attitude of transplanting everything whether or not it is suited to our conditions. This is no good. The other attitude is to use our heads and learn those things which suit our conditions, that is, to absorb whatever experience is useful to us. This is the attitude we should adopt."

Afterword

The 12 lectures contained in this publication were originally given at the Technological Universities of Eindhoven and Twente at the end of 1974. Financed by the Netherlands Universities Foundation for International Cooperation, the course was organized to provide a technically detailed introduction to socially appropriate technology for the growing number of interested people in the Netherlands. It was apparent that in Britain, particularly in the Intermediate Technology Development Group, was to be found the most practical experience in mobilizing scientific and technical knowledge for the developing world. And so it was from there that most of the speakers were drawn.

In each of the two universities the interest and involvement in development affairs falls under the aegis of a Committee for International Cooperation Activities (CICA). These committees were responsible for organizing the course of lectures, and the administration was provided by the Bureau Ontwikkelingssamenwerking (Office of Development Cooperation) at Eindhoven in conjunction with the CICA Secretariaat at Twente. Since both are technological universities, it is natural that a good part of the activities of their CICA committees should be concerned with the relationship of technology to development. Over the past years in the Netherlands there has been growing interest in intermediate technology, and the phrase "socially appropriate technology" has been coined to describe the Dutch approach. At the national level, the activities of university and other groups are coordinated by the newly formed TOOL Foundation (Technische Ontwikkeling Ontwikkenlings Landen—or Technical Development for Developing Countries). The audience for these lectures, then, was drawn not only from the two universities, but also from the membership of TOOL.

As the title implies, each contribution presented here is a lecture; the publication really consists of an edited version of a range of fairly informal talks. Although each contributor has obviously carefully prepared his presentation, he has approached his task very differently than he would if he had been asked to submit a written paper for a collection of essays. I think that the results printed here probably make easier and more enjoyable reading than a series of comprehensive written papers.

Again, because he was talking informally, each speaker gives us an insight into the way he personally views socially appropriate technology—and it is apparent that there are several viewpoints. Although many of the speakers are associated with ITDG, what they have to say does not, of course, represent all of that group's activities, or those of the panels they work with, or of their own personal research of professional interests. So each speaker's contribution is merely an account of the work he is associated with or interested in. It is a general introduction to the more specialized subject, and will hopefully further encourage other technical specialists to expand the potential of technology for the developing world.

Clearly within the time available, each speaker could only outline a small amount of the potential of his own discipline. It would have been impossible to place socially appropriate technology, even less the specialized subject, within a framework of the development "problem" or of current development strategies and institutions. There is a wealth of literature on what has been called "developmentology"; it is hoped, rather, that this publication will provide both an indication of the potential of small-scale technology and a challenge to technologists to realize that potential.

Ina Manni and Hanna Goossen did all the transcription, Kathleen Rutten translated into English the lectures of Ben van Bronckhorst and Ton de Wilde, and the secretaries of BOS typed and retyped the transcripts as they were edited; all of them were working in a foreign language. Bryony Lee typed the final manuscript. To all of them and to the staff of the Reproduction Department go my thanks.

R. J. Congdon
Department of
Industrial Engineering
Technische Hogeschool
Eindhoven, Netherlands

Appropriate Technology Groups

Alternative Energy Resources Organization (AERO)
435 Stapleton Building
Billings, MT 59101

Appropriate Technology International (ATI)
c/o Agency for International Development (AID)
U.S. State Department
Washington, DC 20523

Brace Research Institute
P.O. Box 400
McDonald Campus of McGill University
Saint Anne de Bellevue 809
Quebec, CANADA

Center for Maximum Potential Building Systems
6438 Bee Cave Road
Austin, TX 78746

Conservation Tools and Technology Ltd. (CTT)
P.O. Box 134
Surrey KT2 6PR
ENGLAND

East–West Center
Technology and Development Institute
Honolulu, HI 96822

Ecotope Group
Box 618
Snohomish, WA 98290

Farallones Institute
P.O. Box 700
Point Reyes, CA 94956

Groupe de Recherche sur les Techniques Rurales (GRET)
34 rue Dumont d'Urville
75-775 Paris
Cedex 16, FRANCE

IFOAM
International Appropriate Technology Group
Forschungsinstitut fur biologischin Land bau
Postfach, CH-4104
SWITZERLAND

Independent Power Developers
Bill Delt, President
Box 1467
Noxon, MT 59853

Institute for Local Self-Reliance
1717 18th Street, NW
Washington, DC 20009

(Note: All United States ITDG inquiries should be sent to:

International Scholarly Book Services, Inc.
P.O. Box 555
Forest Grove, OR 97116
USA.)

Intermediate Technology Development Group
9 King Street
London,
ENGLAND

International Rice Research Institute
Agricultural Engineering Department
Manila, PHILIPPINES

197

Minimum Cost Housing Group
School of Architecture
McGill University
Montreal, Quebec
CANADA

National Centre For Alternative
  Technology
Llywngwern Quarry
Pantperthog
Machynlleth 2400
Powys, WALES

National Center for Appropriate
  Technology
Box 3838
Butte, MT 59701

New Alchemy Institute
P.O. Box 432
Woods Hole, MA 02543

Office of Appropriate Technology (OAT)
Box 1677
Sacramento, CA 95808

OUROBORUS
University of Minnesota
Architectural Department
Minneapolis, MN 55455

Rodale Resources
33 East Minor Street
Emmaus, PA 18049

Small Farm Research Institute
Greenwood Farm
Harborside, ME 04642

Small Industry Development Network
Georgia Institute of Technology
Industrial Development Division
Atlanta, GA 30322

Solar Survival
Lea Poisson
Box 119
Harrisville, NH 03450

The Solar Sustenance Project
Bill Yanda
Route 1, Box 107AA
Santa Fe, NM 87501

TECHNOSERVE, Inc.
P.O. Box 409
Greenwich, CT 06830

TOOL Foundation
(Technische Ontwikkeling Ontwikkelings
  Landen)
Mauritskade 61a,
Amsterdam, NETHERLANDS

Total Environmental Action (TEA)
Harrisville, NH 03450

Transnational Network for Appropriate/Al-
  ternative Technologies (TRANET)
c/o William Ellis
7410 Vernon Square Drive
Alexandria, VA 22306

Volunteers for International Technical
  Assistance (VITA)
3706 Rhode Island Avenue
Mount Rainier, MD 20822

Volunteers in Asia, Inc. (VIA)
Box 4543
Stanford, CA 94305

The Walden Foundation
c/o James DeKorne
P.O. Box 5
El Rito, NM 87530

Windworks
Route 3, Box 329
Mukwonago, WI 53149

Zomeworks
P.O. Box 712
Albuquerque, NM 87103

Appropriate Technology Publications

ACORN
Midwest Energy Alternative Network
Governors State University
Park Forest South, IL 60466

Alternate Sources of Energy
Route 2, Box 90-A
Milaca, MN 56353

Co-Evolution Quarterly
Box 428
Sausalito, CA 94965

Doing It!
Box 303
Worthington, OH 43085

The Futurist
World Future Society
P.O. Box 30369
Bethesda Branch
Washington, DC 20014

META Publications
P.O. Box 128H
Marblemount, WA 98267

RAIN Magazine
2270 NW Irving
Portland, OR 97210

*Technical Assistance Information Clearing
House (TAICH)*
200 Park Avenue South
New York, NY 10003

TILTH Newsletter
Box 2382
Olympia, WA 98507

Undercurrents Ltd.
11 Shadwell
Uley
Dursley
Gloucestershire,
ENGLAND

Wind Digest
54468 CR 31
Bristol, IN 46507

Bibliography

Integrative Design Associates, Inc. February 1977. *Appropriate Technology in the United States: An Exploratory Study*. Washington, DC.

Baron, C. G. 1975. Sugar Processing Techniques in India, A. S. Bhalla, ed. *Technology and Employment in Industry*. Geneva: International Labour Office.

Bateman, G. H. 1974. *A Bibliography of Low-Cost Water Technologies*, 3rd ed. London: Intermediate Technology Development Group.

Beale, W. T. 1969. *Free Piston Stirling Engine, Some Model Tests and Simulation*. SAE International Auto, English Conference. Detroit, MI.

Bertholet, C. J. L. 1967. *Het Ontwikkelingsprobleem in Sociaal Perspectief*. Rotterdam: Universitaire Pers.

Bhalla, A. S., ed. 1975. *Technology and Employment in Industry*. Geneva: International Labour Office.

Carruthers, D. 1973. *Impact and Economics of Community Water Supply—A Study of Rural Water Investment in Kenya*. London: Wye College of University of London.

United Nations. 1971. *Climate and House Design*. Sales Number: E.69.IV.11. New York.

Coombs, Philip. *New Paths to Learning*. New York: UNICEF.

Dalton, A. J. P. 1973. *Chemicals from Biological Materials*. London: Intermediate Technology Development Group.

de Bruijn, E. J. 1972. *The Applicability of Modern Methods of Management and Production in the Indonesian Metal Industries*. Twente: Enschede Technische Hogeschool.

Darrow, Ken and Pam, Rick. 1976. *Appropriate Technology Sourcebook*. Stanford, CA: Volunteers in Asia, Inc.

Dickson, D. 1974. *The Politics of Alternative Technology*. New York: Universe Books.

Dumont, R. 1968. *False Start in Africa*. London: Sphere.

Centre D'etudes et d'experimentation du machinisme agricole tropical. 1972. *The Employment of Draught Animals in Agriculture*. Rome: FAO.

FAO. 1961. *Farm Implements for Arid and Tropical Regions*. no. 91. Rome.

National Academy of Sciences. 1973. *Ferrocement Applications in Developing Countries*. Washington, DC.

Harahap, F. August 1972. Teknologi tepat bagi pembangunan industri. *Prisma* 5:58–62.

Humphrey, H. A. November 1909. *An Internal Combustion Pump and Other Applications of a New Principle*. London: Proc. I. Mech. E.

Intermediate Technology Development Group. 1975. *The Iron Foundry—An Industrial Profile*. London.

Jackson, S. 1972. *Economically Appropriate Technologies for Developing Countries, A Survey*. no. 3. Washington, DC: Overseas Development Council.

Jéquier, Nicolas. 1976. *Appropriate Technology: Problems and Promises*. Paris: Development Center of the Organization for Economic Cooperation and Development.

Khan, A. 1974. *Mechanisation Technology for Tropical Agriculture*. Los Banas, Philippines: International Rice Research Institute.

Kline, Green, Donahue, and Stout. 1969. *Agricultural Mechanisation in Equatorial Africa*. Institute of International Agriculture research report no. 6, East Lansing, MI: Michigan State University.

MacCracken, C. D. 1955. *The Solar Powered Thermopump*. Tucson, AZ: Trans. Conference on Solar Energy.

Oregon State University. 1972. *Manual of Pesticide Application Equipment*. Corvallis, OR: International Plant Protection Center.

Intermediate Technology Development Group. 1973. *Manual on Building Construction*. London.

Marsden. 1970. Progressive Technologies. *International Labour Review* 101:475–502.

National Academy of Sciences. 1974. *More Water for Arid Lands—Promising Technologies and Research Opportunities*. Washington, DC.

Needham, J. 1965. *Science and Civilisation in China*. vol. 4, part 2, sect. 27. Mechanical engineering. London: Cambridge University.

———. op cit. vol 4, part 3. Civil engineering and nautics. London: Cambridge University.

Nyerere, Julius. *Education for Self-reliance*. Dar-es-Salaam, Tanzania: Government Printer.

Payne, P. R. August 1974. *A New Steam Engine Cycle*. San Francisco: ICECE Center.

Wilson, S. S. *Pedal Power Progress*. An occasional newssheet from Engineering Laboratory, Parks Road, Oxford, England.

National Center for Appropriate Technology. September 7, 1976. *Proposal for the National Center for Appropriate Technology*. Butte, MT.

Reynolds, G. F. 1975. The Generation of Methane. *Appropriate Technology*. vol. 2, no. 2.

———. 1972. The Ox Plough Revolution. *Chemistry in Britain*. 8:534.

Schumacher, E. F. 1962. Levels of Technology. *Roots of Economic Growth*. Varanasi, India: Gandhian Institute of Studies.

———. 1973. *Small Is Beautiful*. London: Blond and Briggs. In Dutch: 1974. *Hou Het Klein*. Bilthoven: Ambo.

Scheffield, James and Diejomaoh, P. *Non-formal Education in African Development*. New York: African-American Institute.

Intermediate Technology Development Group. 1973. *Simple Designs for Hospital Equipment*. London.

Spence, R. G., ed. 1975. *Lime and Alternative Cements–A Report of a Colloquium*. London.

Intermediate Technology Development Group. 1972. *The Stirling Engine—An Annotated Bibliography*. London.

International Development Research Centre. 1973. *Technology Assessment and Research Priorities for Water Supply and Sanitation in Developing Countries*. Ottawa.

UN Centre for Economic and Social Information. 1973. *The Case for Development*. New York and London: Praeger.

van Bronckhorst, B. 1974. *Development Problems in the Perspective of Technology*. Eindhoven: Technische Hogeschool Eindhoven.

International Bank for Reconstruction and Development. 1974. *Village Water Supply and Sanitation in Less Developed Countries*. Public Utilities research report no. res. 2. Washington, DC.

West, C. 1974. *The Fluidyne Heat Engine*. research report R 6775. Harwell, England: UKAERE.

Whitt, F. G. and Wilson, D. G. 1974. *Bicycling Science—Ergonomics and Mechanics*. Cambridge, MA: MIT Press.

Wilson, S. S. March 1973. Bicycle Technology. *Scientific American*. no. 3, 228:81.

Index